KU-309-151

Learning Modelling with DERIVE

M. Stewart Townend
David C. Pountney
School of Computing and Mathematical Sciences
Liverpool John Moores University

PRENTICE HALL

London New York Toronto Sydney Tokyo Singapore Madrid Mexico City Munich

First published 1995 by
Prentice Hall International (UK) Limited
Campus 400, Maylands Avenue
Hemel Hempstead
Hertfordshire, HP2 7EZ
A division of
Simon & Schuster International Group

© Prentice Hall International (UK) Limited 1995

All rights reserved. No part of this publication may be reproduced, stored in a retrieval system, or transmitted, in any form, or by any means, electronic, mechanical, photocopying, recording or otherwise, without prior permission, in writing, from the publisher.

Printed and bound in Great Britain by
T.J. Press Ltd, Padstow, Cornwall

Library of Congress Cataloging-in-Publication Data

Available from the publisher

British Library Cataloguing in Publication Data

A catalogue record for this book is available
from the British Library

ISBN 0-13-190521-X

1 2 3 4 5 99 98 97 96 95

Learning Modelling with DERIVE [illegible]

LIVERPOOL JMU LIBRARY
3 1111 00733 3048

Contents

Series Preface

This series is designed to encourage the teaching and learning of various courses in mathematics using the symbolic algebra package DERIVE. Each text in the series will include the 'standard theory' appropriate to the subject, worked examples, exercises and *DERIVE Activities* so that students have a sound conceptual base on which to build. It is through the DERIVE activities that the learning of mathematics through investigations and discussion is encouraged. The aims of each DERIVE activity are to provide an investigation to introduce a new topic and to introduce the commands for using DERIVE as a manipulator in mathematical problem solving. Mathematics teaching and learning is changing as new technology becomes more readily available and these texts are written to support the new methods.

John Berry
Centre for Teaching Mathematics
University of Plymouth

Preface

Mathematical modelling now enjoys a central role in many mathematics degree courses in universities, having risen to this position during the past two decades in response *inter alia* to the demands of employers who require graduates who can analyze and solve problems and communicate their conclusions to others.

In the past five to ten years the teaching of undergraduate mathematics has been transformed from a 'chalk and talk' approach to a much more interactive, laboratory-based approach centred on the use of mathematical software, notably computer algebra packages such as DERIVE®. The students' consequent release from long, often tedious mathematical manipulations and calculations has increased their mathematical enthusiasm and encouraged them to consider the applications of their mathematics. The existence of a readily available package such as DERIVE raises an interesting pedagogical question as to whether students should also demonstrate manual ability within the course content – a question which is addressed in the companion text in this series *Learning Mathematics Through DERIVE.*

Our text integrates DERIVE into the learning of mathematical modelling by using it to assist in the solution and subsequent investigation of the mathematical models derived. As mentioned above, the modellers' release from much of the labour involved in producing the solution leaves them free, and more inclined, to address the interpretation of the solution and the development of revisions to their model. Moreover, the support provided by DERIVE means that the model can be more realistic *ab initio*.

The first chapter looks at modelling from the perspective of both the student and the tutor and is used to set the scene, illustrate a modelling methodology via a worked example and address the issue of assessment. The modelling problems presented in the text are grouped according to the typical syllabus content of first- and second-year undergraduate mathematics courses. Some of the problems are adaptations of previously published problems. We urge the reader to investigate these references too. For example in Chapter 2, which looks at models of a trigonometric or geometric nature, we have included the right hand drive version of the well known vehicle squeeze problem associated with the turning of large vehicles. The solution mirrors closely the left hand drive version presented in detail by E.A. Bender in An Introduction to Mathematical Modelling (Wiley, 1978). Chapter 3 addresses algebraic models, while Chapter 4 provides the first links with calculus via problems in optimization, relevant linear programming techniques also being presented. Chapter 5 considers the important topic of statistical modelling. Chapters 6 and 7 deal with continuous and discrete models, with due consideration given to a rationale for deciding which approach is appropriate for a given problem. We debated for some time as to whether to include Chapter 8, which introduces the concept of dimensional analysis. On the one hand, the mathematics involved scarcely warrants any mathematical assistance and yet the method represents such a powerful modelling tool that we felt it should be included in a book about mathematical modelling. As you can see, we finally decided on its inclusion.

Chapter 9 contains a miscellany of sixteen modelling problem outlines. We have deliberately collected them here rather than distribute them throughout the text so that the modeller is not 'led' by the title of the chapter in which the problem is located. After all, one of the problems facing a modeller is to decide what mathematics is likely to be appropriate.

The final chapter addresses the issue of changes in the mathematical skills base of entrants to higher education. This issue is one of major concern to mathematics educators and course designers both currently and for the foreseeable future . It focuses on reinforcing the understanding of the behaviour of selected mathematical functions and their utility in mathematical modelling. Examples are provided which enable the tutor to develop these links either by exposition or by student-led investigation.

Appropriate DERIVE keystrokes, algebraic and graphical screen dumps (obtained using DERIVE Version 2.58) are presented within the solutions developed in the text to provide guidance for the reader. Suggestions for further DERIVE activities are also given throughout. Some take the form of structured directed activities, while others are open-ended investigations.

A modest level of familiarity with DERIVE is assumed at the outset. Commands with which you should be familiar are

Author, Calculus, Declare, soLve,
Manage, Plot, Simplify, Substitute,
Vector, approXimate

For additional information about these, other specific DERIVE functions (such as FIT, etc.) and the various utility files used frequently throughout the text, the reader is directed to either the *DERIVE User Manual* or the HELP facility available within DERIVE. A complete list of the more advanced DERIVE commands and utility files used in this text is given on page 241.

You may wonder why we decided to link mathematical modelling explicitly with DERIVE ? We did so because, in our opinion, no other computer algebra package currently available possesses both the mathematical power and easy access of DERIVE. In our DERIVE teaching laboratories, we regularly encounter students who, having started on a directed piece of work, will within an hour or so have branched out into an investigation of their own – truly easy-access.

In addition to the various references presented at the end of each chapter, the authors would like to thank Dr Richard Brown (Liverpool John Moores University) for the lizard data of Chapter 5, Dr Graham Raggett (Sheffield Hallam University) for first acquainting the authors with the Indian population policy statement of Chapter 6 and the unknown driver of an extremely long vehicle for dramatically bringing the right turn swing problem of Chapter 2 to the attention of one of the authors. Thanks are also due to colleagues and students at Liverpool John Moores University who contributed suggestions and have acted as 'guinea pigs' for some of the modelling problems found in this text.

We would also like to thank David Stoutemyer and Albert Rich, the fathers of DERIVE, for several helpful discussions during the writing of this text. *Mahalo nui loa* !

A special thank you must go to Keith Campbell both for his patience and his meticulous production of the camera ready copy of this text.

Finally, we must thank our respective wives and families without whose support none of this text would ever have been written.

We hope that you will enjoy your mathematical modelling with DERIVE.

Stewart Townend
David Pountney

School of Computing and Mathematical Sciences
Liverpool John Moores University
Byrom Street
Liverpool L3 3AF, UK

1

Mathematical Modelling

1.1 INTRODUCTION

Mathematical modelling is now an established component of many undergraduate mathematics degree courses, a position which it has attained over the past two decades or so in response to comments from both employers and academics that while students and graduates are quite good at solving mathematical problems they are very poor at formulating the problem in the first place and in communicating the results to others.

The then polytechnics, and some universities, were quick to address these issues and to develop and launch modules/courses which embraced both mathematical modelling and problem solving. More recently, changes have also taken place within the school mathematics curriculum so that school leavers have now experienced an open-ended, investigative approach to problem solving. This has further impinged on the syllabus content within the higher education sector and strengthened the position occupied by mathematical modelling.

It is interesting to note that many of the job advertisements targeted at new graduates specifically mention the need for skills in problem solving, ability to work as part of a team and communication skills – precisely the skills which modelling courses seek to develop, and which the authors wish to enhance with this text.

Computer algebra packages such as DERIVE are now widely available and they free students from a considerable amount of the mathematical drudgery associated with the solution of a mathematical model, thereby enabling them to spend more time on interpretation of the solution, sensitivity analysis, model refinement and investigation of the 'What if...?' type of questions which they should consider.

What is a Mathematical Model?

A possible definition is as follows:

> **A mathematical model is the translation of a real world problem into its mathematically formulated analog. This is then solved and the resulting solution translated back into the original context.**

It is more than likely that when the first solution is interpreted it will not be entirely satisfactory. For example, it might not identify any known special cases or it might not adequately describe the real world problem at all. Consequently, the mathematical modeller needs to have an open mind and must be prepared to refine the original model and go through the above process again, i.e. modelling should be seen as an iterative process.

A Modelling Methodology

The open-ended nature of mathematical modelling exercises is the antithesis of the one problem–one solution experience of most undergraduate mathematicians (and others, too). As already mentioned, this will change in the future as entrants to the university sector will already have had some problem solving experience at school.

Nevertheless, the authors consider it appropriate to provide a framework around which new practitioners can develop their modelling skills.

The methodology is illustrated in Figure 1.1. It is based upon the Open University's seven-box approach [1] and reflects an approach which is widely used in the teaching of mathematical modelling.

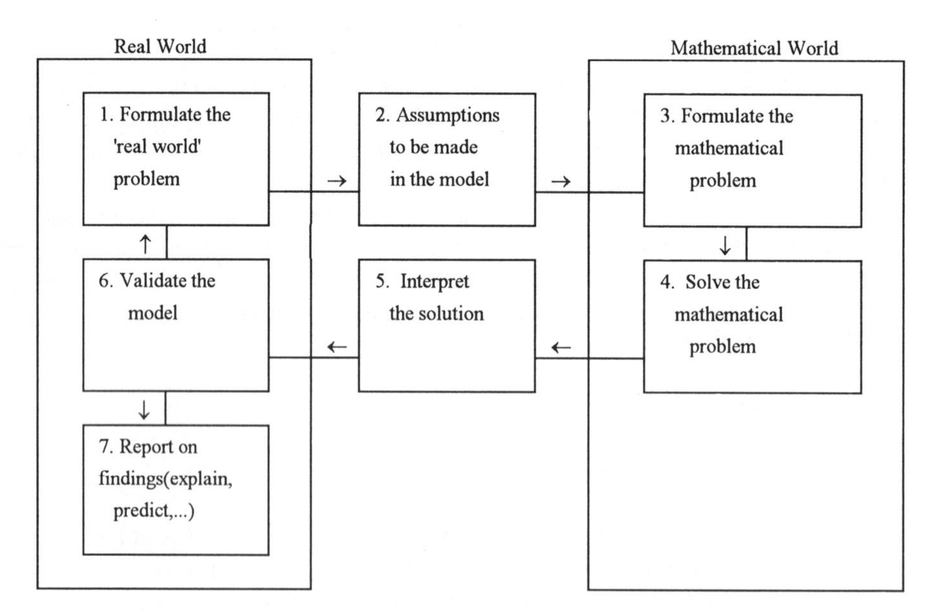

Figure 1.1 A modelling methodology – the modelling loop

With the class divided into small groups (see the next section for a discussion of the rationale for this) the problem is discussed and agreed with the client, who might be a role-playing tutor. Hopefully, an initial brain-storming session will then produce an extensive list of features and assumptions which will need to be pruned in order to produce a tractable initial model (one of the 'golden rules' of modelling is 'Keep it simple – at least initially'). One way in which the pruning can be effected is for the group to agree a ranking for each feature (a possible scale could be 1 = crucial through 4 = trivial/unimportant) and then to formulate an initial model based on the features and assumptions considered to be crucial. This completes the stages identified in boxes 1 and 2 of Figure 1.1.

In box 3 we enter the mathematical world. Here we define the various variables to be used in the model and then use them to formulate the assumptions mathematically. This might result in an expression to be optimized, an integral to be evaluated, a differential equation to be solved, and so on. In box 4 the mathematics required to obtain the solution is carried out. On completion of this stage the modellers will have a solution either as a closed-form mathematical expression or perhaps a table of numerical data.

In box 5 the crucial question 'What does our solution tell us?' is addressed. At this stage the modellers should be asking such questions as 'Do we believe this answer?', 'What happens if...?', etc. – in short they should be challenging their solution. It often happens that the answers to these questions are less than acceptable and this implies that the original model is flawed in some way. At this point it is necessary to revise the original model, perhaps to also include those features of rank 2, and go round the modelling loop of boxes 1–6 again. This iterative process will hopefully lead to a solution which the modellers consider to be acceptable.

At the box 6 stage, the solution is discussed with the client and checked with known data or validated by experiment, if this is possible/feasible. Finally, a decision is made as to whether any further refinements of the model are considered necessary/cost effective. In box 7, the findings are reported to the client in an agreed format.

The various stages identified here and in Figure 1.1 are highlighted in the worked example presented later in this chapter. It should be emphasized that in practice the modelling loop is unlikely to be followed rigidly. Certainly, after the initial model has been formulated, any revisions are likely to be arrived at by 'jumping' backwards and forwards between some of the boxes and omitting some of the intermediate ones. This reflects the fact that modelling is not as structured an activity as, say, the development of the solutions of mathematical technique-bashing exercises. The modelling methodology responds to sudden 'flashes of inspiration' which then leapfrog the modellers a considerable way round the modelling loop. Occasionally, a model may prove inappropriate even after many revisions have been made and a completely fresh start will be needed. Modelling is an iterative process and not all such processes converge, especially if the first iterate is inappropriate!

A Teaching Strategy

It should not have escaped your notice that earlier we referred to groups and modellers (plural). This was not a typographical error. The previous section refers, at the box 1 and 2 stages, to an initial brain-storming session. Obviously, for this to be productive it must involve a group of people. Ideas can then be exchanged, criticized and defended until, ultimately, a consensus point of view is established as the basis for the initial model. Constructive group dynamics is central to the success of this approach. Too few people in a group and there will not be enough interaction, too many and it is unlikely to reach many conclusions. Experience suggests that a group size of about four students is optimal for modelling activities and, further, that in the classroom one tutor can then satisfactorily monitor the progress of about six groups.

The classroom used should contain free-standing furniture which can easily be arranged for group discussions. Likewise, dedicated computing equipment and software should be readily available but, failing this, proximity to a computer laboratory is a minimal requirement.

As for getting started, the following strategy is adopted by the authors. First, divide the class into groups (either tutor selected or student selected). These groups may be composed of students of mixed or ranked ability (if tutor selected) or random ability (if student selected). There is a case to be made for either [2]. Next, provide the class with a selection of problem statements (these are each typically a one- or two-paragraph written statement of the problem to be modelled and solved) and give the groups about half an hour to decide which problem they wish to tackle and then to begin to formulate some ideas. Thereafter, the groups can get on with their chosen problem using the methodology suggested earlier. During this time the tutor circulates among the groups, 'eavesdropping' on their discussions, assessing the input of individual group members and providing guidance/arbitration, as required, while being careful not to lead the discussions.

Most of the case studies in this book have been successfully used in first- and second-year undergraduate modelling courses and are able to be completed in about nine hours of group activity plus any private study time which the students may wish to add.

Assessment of Modelling Reports

A mathematical modelling course requires a student to exhibit a wide range of skills, as follows:

- problem formulation,
- initiative,
- mathematical ability (to obtain a solution),
- discussion of results,
- revisions,
- clarity and brevity,
- communication skills (verbal or written),
- ability to work in a team.

Any assessment scheme must address this multifaceted aspect and many have been developed for the assessment of verbal and written reports, see, for example, Table 1.1.

Table 1.1 Construction of written report

Section	Contents	Mark
Summary	A clear statement of the problem. Starting point defined. Conclusions stated.	5
Formulation	Simplifying assumptions clearly stated. Feature list and pruning presented with reasons for omitting pruned features. Written statement of the model to be used.	10
Data	Sources acknowledged. Are they appropriate? Details of any experimental work. Clarity of presentation.	10
Initial model	All variables defined. Formulation of mathematical relations and equations. Solution and its interpretation. Use of data to test results. Critical appraisal.	45
Revised model	Description of revision on basis of above criticism. Solution of revised model and its interpretation. Suggestions for further work.	30
	TOTAL	100

Following the increasing use of poster sessions at conferences, several universities are now experimenting with their use as a means of assessing undergraduate modelling activities, see, for example, [3].

Ideally, assessment should include both a written report and a verbal or a poster presentation (with attendance for a defence in the latter case), thereby assessing the students' written and oral powers of communication. Such reports are usually submitted from each group rather than individually and it is therefore necessary to provide some discriminatory aspect to the assessment of each individual within a group. Various schemes exist, see [4]. The authors use a scheme in which each complete group openly negotiates a factor (≤ 1) for each individual in that group. This factor is then applied to the mark awarded to the report to produce a mark which reflects the effort of each individual member of that group. Most recently, see [5], another form of peer assessment (in which members of one group quantitatively assess the performance of other groups) is being trialled in some institutions, the result being used as an additional input to the overall tutor assessment.

The creative, open-ended nature of mathematical modelling suggests that the traditional, timed, closed-book examination is not appropriate if used with previously unseen problem statements. However, timed, written assessment can be used if it is directed at critical mathematical discussion of the oral/written/poster presentations made by class members during the course. This provides a further discriminatory element to the assessment.

At the beginning of this section, we listed some attributes of a modelling course in which students might expect to be assessed. Some of these might be considered to be more important than others and can therefore be more heavily weighted. Once the component marks have been assigned they are, traditionally, summed to provide an overall mark for the presentation. The 'participation factor' is then applied to this mark to produce a mark for each individual within a group. It can be argued that this summation model is flawed in the sense that a high overall mark should imply a good performance in each facet assessed, whereas it can also include the case where a weak performance in one category is masked by an excellent performance in another. This phenomenon can be overcome by generating a combined mark based on a geometric mean rather than an arithmetic mean, see Table 1.2. For example, with three component marks x, y and z respectively weighted 2, 2 and 1, the arithmetic mean A and the geometric mean P are given by:

$$A = (2x + 2y + z)/5$$

$$P = \sqrt[5]{x^2 y^2 z}$$

Two things are immediately apparent from Table 1.2:

(a) A disastrous performance in any component has a disastrous overall effect.
(b) Whichever model is used there is little difference in the result if all the component marks are reasonable.

Investigation of the two schemes outlined above makes an interesting article to present to a class at the start of a modelling course to get them talking about mathematics and to realize that problems can have more than one solution!

Table 1.2 Marking scheme comparison

Component Marks $x,\ y,\ z$	A	P
50, 50, 50	50	50
50, 60, 50	58	57.8
32, 50, 40	40.8	40
32, 32, 0	25.6	0
32, 1, 32	19.6	8
90, 10, 30	46	30

In oral presentations the students usually make use of an overhead projector, and members of the groups often each take responsibility for a part of the presentation. Criteria similar to those listed in Table 1.1 can again be used for assessment but the marking scheme needs to be adapted to include assessment of communication skills. The sort of things to look for are as follows:

- Did the speaker talk about the problem or just read from notes?
- Was the speaker lucid, brief and able to be heard?
- Was relevant technology (e.g. overhead projector or live interaction with software) used appropriately and effectively?
- Was the speaker confident (i.e. was there eye-contact with the audience?)?

From the tutor's point of view, one of the difficulties which arises if there are many oral presentations is the difficulty in recalling earlier presentations accurately for comparison with later ones. This indicates one of the advantages of the poster session as lasting evidence is available at the end of the assessment session. The students' communication and interpersonal skills can still be assessed during the poster defence.

A poster should communicate the main ideas visually. It should be clear, concise, uncluttered and attractive. Diagrams and pictures are helpful in this latter respect, but they are space filling. The limited space imposed by posters encourages students to be brief and concentrate on the main message, while at the same time producing a result which is aesthetically pleasing. It follows, therefore, that content and design are the main features of importance with a poster and students obviously need specific guidance as to what to include. The following lists indicate such features, see [3].

Content:

the problem statement
relevant information (including modelling assumptions)
outline of solution
conclusions

Design:

visually attractive
large print
logical layout
concise
clear headings
use of pictures/diagrams

All the posters are subsequently displayed at a poster session in which each student group discusses and defends its poster with other students and the tutor. This informal atmosphere is less intimidating to some students than a formal oral presentation. If desired it can be made into a feature of a department's activities by inviting colleagues and providing refreshments, just as in a poster session at a conference, see [3].

Berry and Houston [3] suggest the following list of assessment criteria to which the present authors have appended a marking scheme.

Marking Criteria

Criterion		Mark
problem statement clear and concise		40%
outline of problem	– describes the model	
	– states the mathematical problem	
	– reports on the solution	
report conclusions		
poster design is logical with bold headings, appropriate use of illustrations and is aesthetically pleasing overall		40%
discussion/defence demonstrates understanding of the project		20%

Students who have experienced poster presentations report that they enjoyed the challenge and sense of occasion. It should be noted that at, the time of writing, this approach is relatively new to mathematics education and our apportioning of marks will probably change in the light of operational experience.

We conclude this introductory chapter with a worked example in order to illustrate the various stages of the modelling methodology presented in Fig. 1.1 and to provide readers with a framework which can then be used to develop their subsequent modelling activities.

1.2 PLACE KICKING IN RUGBY FOOTBALL

The Problem

In the game of rugby football, once a try has been scored, the scoring team has the opportunity to take a conversion kick. What is the best place P from which to take the kick if the try is scored at the point T shown in Fig. 1.2?

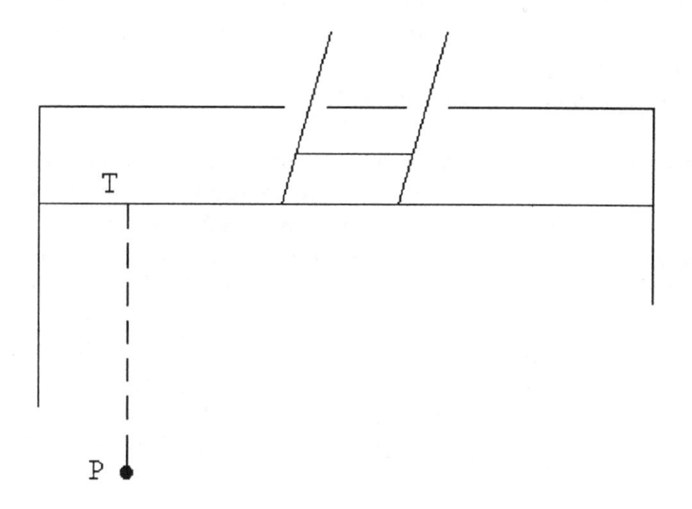

Figure 1.2 The optimum position P for a conversion kick

Setting up a Model

Before attempting to draw up a feature list it may be necessary for the modelling group to consult an 'expert' (the tutor?) to establish that the rules of the game state that the kick may be taken from any point P on a straight line through T drawn parallel to the touchline. (Consultation with an 'expert' is often required when formulating the real world problem in box 1.)

Constructing a feature list is the box 2 group brain-storming session referred to earlier and is likely to lead to a list such as the following:

		rank
1.	distance from P to the posts	1
2.	horizontal distance from T to the posts	1
3.	distance between the posts	1
4.	wind direction	2
5.	kicking style and run-up adopted	3
6.	kicker's ability to reach the posts and clear the cross-bar	4
7.	initial velocity of ball	4
8.	angle of projection of ball	4

The above is the list actually compiled by one group of the authors' students. It is reproduced here as it illustrates that on reflection by the group they may well find that some of their features are implicit in other features (e.g. 7 and 8 are implicit within 6). In order to prune this list the features were then ranked in importance (earlier we suggested the scale 1 = crucial through 4 = trivial/unimportant); the individual rankings are indicated in the right hand column. Feature 6 has been ranked as a 4 on the assumption that for the conversion kick the captain would only consider using players who were able to reach the posts!

The 'thinking' behind their ranking was as follows:

feature 1 : obviously of critical importance, the kicker would not attempt the kick from a position which was so remote from the posts that the ball wouldn't even reach them!

features 2 and 3 : both of critical importance as they each have an effect on the angle subtended at the posts by the point from which the ball is kicked.

feature 4 : this was ranked as 2 to indicate that while it is believed to have a significant effect on the problem, its inclusion could complicate the initial model significantly, whereas the modelling strategy advocated is to 'keep it simple', at least initially.

feature 5 : this was included in the list as many teams have left- and right-footed kickers to cater for kicks from either side of the pitch.

Individual kickers also adopt either a straight run-up or a curved run-up. Whilst the existence of these subtleties is acknowledged, the basic problem is the same for all – to get the ball between the uprights and over the cross-bar.

features 6,7 and 8 : these are all related to the precision and ability of the kicker and obviously the team captain would not select an unskilled kicker. Consequently, they have each been ranked as being trivial.

Pruning has thus reduced the feature list to the following, more manageable, set of three features:

- distance from P to the posts,
- horizontal distance from T to the posts,
- distance between the posts.

At this stage it is customary to report the reason(s) for the omission of features, thus

- Wind direction – obviously this has an effect. It is initially omitted in order to keep the model simple, but could be incorporated in a more refined model.
- Kicking style and run-up – provided the kicker is sufficiently accurate, the style is secondary.
- Ability to reach the posts – obviously only a player who could kick far enough would be used.

Box 2 has now been completed and we move across into the mathematical world of box 3.

Formulate a Mathematical Model

Eventually, group discussion is likely to establish that, assuming that the kicker can reach the posts and get the ball over the cross-bar, then the best position for P is that which optimizes the angle subtended by the posts at P (see Fig. 1.3).

The authors recall one group who were adamant that this subtended angle continued to increase as P moved down the pitch along the line TP. It was not until they were encouraged to produce some scale drawings that they accepted that the subtended angle attained a maximum value at some specific position of P. The whole class was quite surprised at the amount of insight provided by such a simple piece of geometry, demonstrating the value of 'keeping it simple'.

With the basis of the mathematical model now established, the next step is to define appropriate variables and then to develop the model in terms of this notation. Referring to Fig. 1.3, an origin of coordinates is chosen at the centre of the posts, with respect to which we have

coordinates of T : $(x,0)$,
coordinates of P : (x,y) ,
width between R and Q the foot of each post = $2d$,
where all distances are measured in metres.

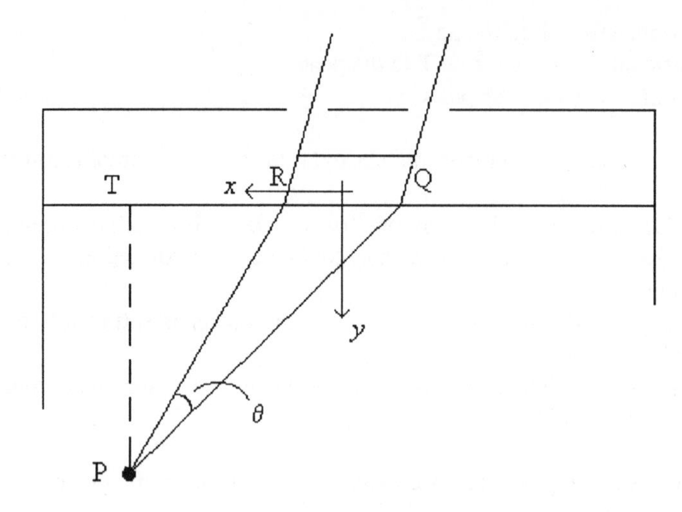

Figure 1.3 The angle subtended at P by the posts

The proposed mathematical model is to maximize the angle θ.
Letting $T\hat{P}R = \alpha$ and $T\hat{P}Q = \beta$, then

$$\theta = \beta - \alpha .$$

Now

$$\tan\alpha = \frac{x-d}{y} \quad \text{and} \quad \tan\beta = \frac{x+d}{y}$$

so that

$$\theta = \operatorname{atan}\left(\frac{x+d}{y}\right) - \operatorname{atan}\left(\frac{x-d}{y}\right)$$

Solution of the Mathematical Model

Discussion will establish that of the variables involved (x,d and y) it is appropriate to maximize θ with respect to y as this is the only variable under the control of the kicker (x is under the control of the try scorer).

The following DERIVE commands lead to the solution to this maximization problem. First, Author the expression for θ. Then select Calculus, Differentiate and carry out one differentiation with respect to y to produce

$$\frac{2d\left(x^2-y^2-d^2\right)}{\left(x^2+2dx+y^2+d^2\right)\left(x^2-2dx+y^2+d^2\right)} \qquad (1.1)$$

The reader should then be able to deduce that the solution of this maximization problem is a hyperbola given by

$$x^2-y^2 = d^2 \qquad (1.2)$$

If DERIVE is used to soLve the equation obtained by setting (1.1) equal to zero, the two expected solutions

$$y = +\sqrt{x^2-d^2} \quad \text{and} \quad y = -\sqrt{x^2-d^2}$$

are obtained together with the spurious solution $y = 1/0$. Consideration of the actual problem shows that of the two solutions the appropriate one is the positive root

$$y = \sqrt{x^2-d^2}\,.$$

Interpretation

This completes box 4. Box 5 concentrates on the interpretation of the solution and can be greatly facilitated using DERIVE's graphics capability.

Graphical output of the solution curve is obtained by first using the Manage and Substitute commands to insert the value of d ($d = 2.8$ m) and then using Plot and Overlay to obtain the graph. You will need to experiment with the Scale and Centre facilities in order to produce an acceptable curve (to aid interpretation you should keep the x and y scales equal, otherwise the curve will be distorted – try x and y scales each set at 2). To summarize the interpretation phase, we see that the angle subtended at the posts by the kicker is maximized if, for a try scored at $(x,0)$, the kicker stands at the point (x,y) where

$$y = \sqrt{x^2-d^2}$$

The graphical output shows that out towards the touchlines the arms of the hyperbola are fairly straight, see Fig. 1.4. This agrees with the fact that the asymptotes to the curve represented by equation (1.2) are given by the solutions of

$$x^2-y^2 = 0$$

i.e. by the two straight lines $y = \pm x$.

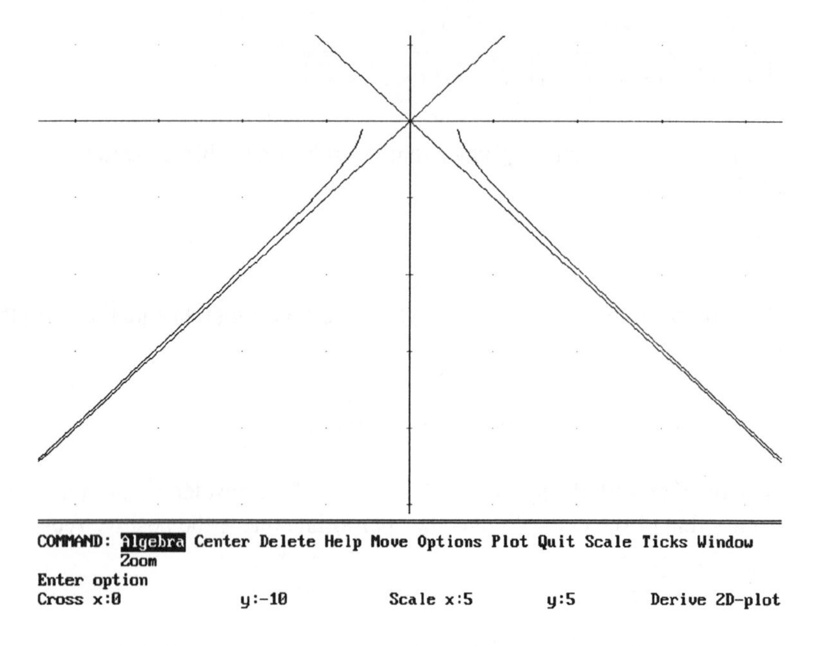

Figure 1.4 The optimal conversion kick hyperbola and its asymptotes

Validation

Finally (box 7), the group could report its findings either verbally or in a written report as follows:

> If the try is scored well out towards the touchlines a good rule of thumb for locating the best place for the conversion kick is to walk back from T, parallel to the touchline, until you reach a point which subtends an angle of 45° between you, the mid-point of the posts and the touchline.

Discussions with rugby players have validated this solution (box 6). The above conclusion is used as a rule of thumb by players and has been arrived at on the basis of their experience.

Mathematically, the solution represented by (1.2) is only valid if $x > d$. So what happens if $x < d$? What does this mean? If $x < d$ then the try has been scored between the posts. In this case the kicking strategy is different. The kick can be taken from in front of the posts (no difficulties with subtended angles), the only requirement being that the kicker is far enough back to be able to clear the cross-bar. This means that our original features 7 and 8 are now important and this special case could be modelled as a projectile problem, see Fig. 1.5.

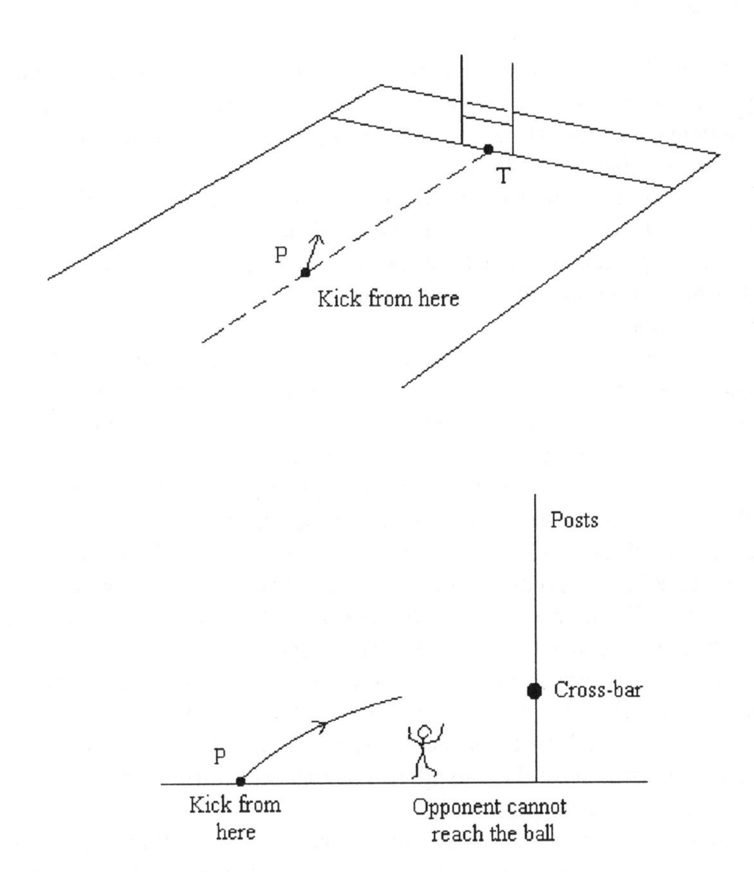

Figure 1.5 Conversion kick for a try scored between the posts

For this special case we also need to bear in mind that the opposing team is permitted to attempt to charge the ball down by running out from between the posts as the kick is taken and trying to catch the ball. This has obvious implications for the angle of projection.

Without discussion of the $x < d$ special case the solution would be considered incomplete. One of the goals of any modelling exercise should be to discover any special cases and address them too.

Further Work

If you wish to develop this model further, the next refinement might be to include wind effects. Try it!

1.3 EPILOGUE

This chapter has served to introduce the rationale for mathematical modelling, a methodology, a teaching strategy, some thoughts on assessment using written reports, oral presentations and posters, a 'worked example' and now concludes with some philosophical points about modelling.

Each section contains pointers designed to help both the modelling tutor with little experience and the novice undergraduate modeller. Not all tutors are comfortable with the group approach to modelling as they consider that there is a risk that 'passengers' within a group may receive unjust rewards. This has been addressed explicitly in the chapter.

Next, a few words to the novice student modeller. One of the potential problems with group work is that of task management within the project. There is no need for everyone in the group to complete every stage of the task – apportion the labour but make sure that there is time for everyone to report back to everyone else so that, ultimately, all members of the group have a complete picture of the problem and its solution. What happens if a group member is absent? Does someone else in the group have a copy of the absentee's contribution so that progress can be continued? What happens if someone in the group is not 'pulling their weight'? It shouldn't happen of course but occasionally there are personality clashes. If negotiations within the group fail, then arbitration will have to be provided by the tutor.

Finally, a few words to the novice modelling tutor. Care needs to be exercised not to lead the students too much, particularly when they seek your advice. Without wasting their time too much let the students experience a 'modelling blind alley'; it is a useful reflective experience for them to appreciate that real world problems don't always slot into place neatly, at least not initially.

While presentation skills are obviously important in writing or presenting any report, high quality word processing and graphics output should not deflect you from the fact that you are primarily assessing the students' modelling ability.

Positively the last words on the topic – modelling is a creative, dynamic and interactive activity. Talk to your fellow group members, defend your point of view and learn to apply your mathematical ability (supported by appropriate software) to develop solutions to real world problems. Problem solving is a highly transferable skill and is valued accordingly in the employment market.

References

[1] Open University Course MST204 Mathematical Models and Methods.

[2] Berry J. and O'Shea T., 'Assessing mathematical modelling', *Int. J. Math. Educ. Sci. Technol.*, **13**(6) (1982).

[3] Berry J. and Houston K., 'Students using posters as a means of assessment', *Educ. Studies in Mathematics* (to appear).

[4] Goldfinch J.M. and Raeside R., 'Development of a peer assessment technique for obtaining individual marks on a group project', *Assessment and Evaluation in Higher Education*, **15**(3) (1990).

[5] Houston K. et al., 'Self and peer assessment', *Proceedings of Undergraduate Mathematics Teaching Conference, Shell Centre for Mathematical Education, University of Nottingham* (1993).

2

Trigonometric Models

2.1 INTRODUCTION

In this chapter we shall investigate some modelling problems whose solutions can be obtained using geometric or trigonometric arguments. We have already seen one such model in Section 1.2 of Chapter 1, where we investigated the optimal position from which to take a conversion kick in rugby football.

We begin the chapter by considering a traffic problem familiar to many drivers.

2.2 THE RIGHT TURN SWING

The Problem

Fig. 2.1 shows a large vehicle positioned at a road junction from which it is going to make a right turn. Imagine that you are in the car C in the inside lane and close to the rear corner A of the large vehicle. As the large vehicle moves off, the point A will move towards your vehicle. The traffic in front of you has not moved and you notice the point A getting closer to the side of your car. Should you be worried? How close will it get?

Setting up a Model

The first step in setting up a model is to identify the important features. The list below represents the authors' thoughts:

- the wheelbase of the large vehicle (w metres) ,
- the width of the large vehicle ($2t$ m) ,
- the rear overhang of the large vehicle (h m) ,
- the angle through which the front wheels of the large vehicle have turned (α degrees),
- the angle through which the vehicle body has turned (β degrees) ,
- the speed at which the vehicle is turning (v m s^{-1}).

Having considered the features we now turn our attention to developing a set of modelling assumptions. We shall assume the following:

- The wheels do not slide sideways during turning.
- v represents the speed of the centre of the front axle.

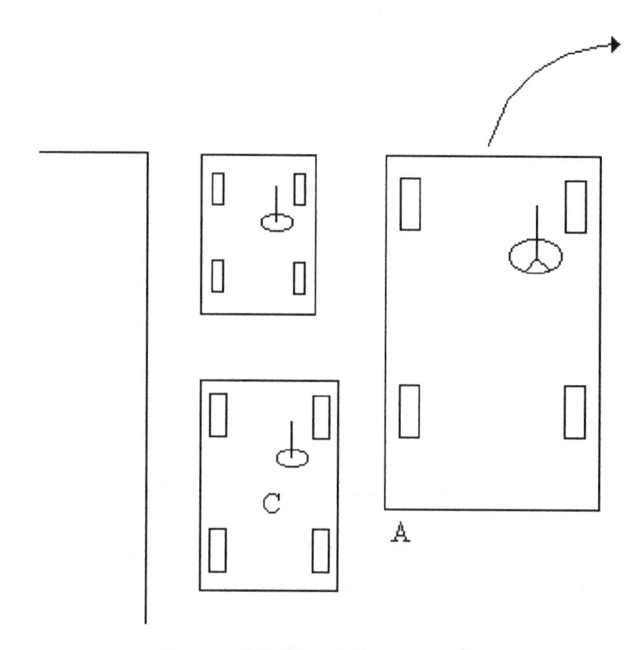

Figure 2.1 The right turn swing

It is important to be so precise about specifying the point where the vehicle's speed is measured since not all points of the vehicle move at the same speed as it turns.

With our variables defined and the modelling assumptions established we are ready for the next stage.

Formulate a Mathematical Model

We need to develop a model to describe the motion of the turning vehicle and then to investigate the lateral shift of the point A as the vehicle moves.

Fig. 2.2 shows the vehicle at some time after the turn has been initiated. During a time interval dt (seconds) of the turning phase the mid-point P of the front axle moves a distance $v\,dt$ parallel to the direction in which the front tyres point. This may be resolved into displacements of

$v\,dt \sin\alpha$ perpendicular to PQ and
$v\,dt \cos\alpha$ parallel to PQ.

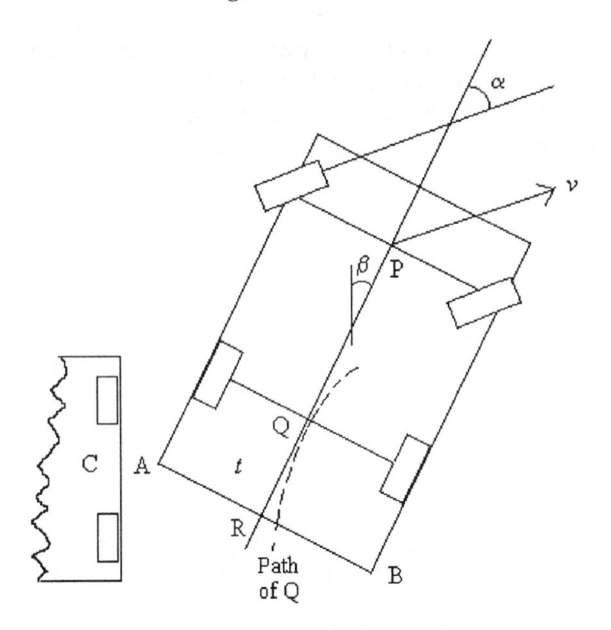

Figure 2.2 The geometry of the turn

Measured with respect to Q, the mid-point of the rear axle, the displacement of P may be expressed in terms of $d\beta$ (the change in β in the time dt) as

a displacement $w\ d\beta$ perpendicular to PQ

while the displacement parallel to PQ depends on the locus of Q. Thus we may write

$$w\ d\beta = v\ dt\ \sin\alpha \tag{2.1}$$

Throughout the remainder of the development of the mathematical model the reader may find Fig. 2.3 helpful in establishing the various results.

If the turn takes a total time t seconds, show that the total displacement of P to the right (measured from the initial position of the centreline of the vehicle) is given by

$$x = \int_0^t v \sin(\alpha + \beta)\ dt$$

and that combining this with equation (2.1) gives

$$x = w \int_0^\beta \frac{\sin(\alpha + \beta)}{\sin\alpha} d\beta \tag{2.2}$$

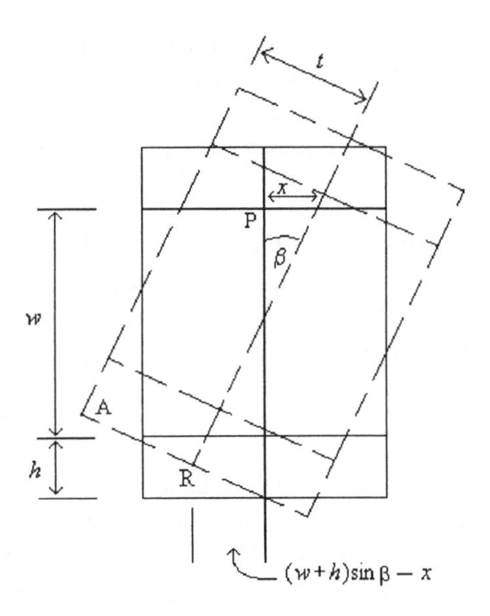

Figure 2.3 Analysis of the left shift of A

Show, further, that the displacement of R to the left (measured from the initial position of the vehicle's centreline) is

$$(w+h)\sin\beta - x$$

and that the displacement of A to the left (measured with respect to the initial position of the centreline of the vehicle) is given by the following function of α and β:

$$\begin{aligned} f(\alpha,\beta) &= (w+h)\sin\beta - x \ + \ t\cos\beta - t \\ &= (w+h)\sin\beta + t\cos\beta - t - w\int_0^\beta \frac{\sin(\alpha+\beta)}{\sin\alpha}\,d\beta \end{aligned} \qquad (2.3)$$

This completes the geometric and trigonometric analysis of the problem and we now turn our attention to finding the maximum value of f.

Solution of the Mathematical Model

Since we are interested in the maximum value of f we shall now be interested in a calculus approach – but should we differentiate f with respect to α or β or both? Clearly, β varies during the turn, so that is one obvious candidate for differentiation, but what about α?

Remember that α denotes the angle through which the front wheels have been turned. A reasonable assumption to make at this point would be to keep α constant and then regard f purely as a function of β. This represents good modelling practice – it is reasonable (it corresponds to the case where the driver keeps the wheels turned at a constant angle) and it simplifies the model. We shall use this assumption in our solution.

An alternative assumption, left for the reader to investigate, is to assume that $\alpha + \beta$ is constant – this corresponds to the driver keeping the front wheels pointing in a fixed direction.

With α constant, the function in equation (2.3) becomes a function of a single variable β whose maximum value can be determined in the usual way using DERIVE. In this exercise the procedure becomes quite involved and so the various stages are presented in Fig. 2.4.

Expression 4 shows the Authored expression for f. Use Manage and Trigonometry with Direction in Expand mode, to deal with the sin $(\alpha + \beta)$ term, and Simplify to obtain expression 5. Next use Calculus and Differentiate to obtain the first derivative of f with respect to β (expression 7) and finally soLve this for the optimal value of β, see expression 8.

The movement of the point A to the left is then given by expression 5 but with β replaced by its optimal value and Simplified (!) to give expression 10.

The algebra of the solution is becoming extremely unwieldy and we can simplify our working considerably by inserting some specific data values. The authors' measurements of a bus gave $w = 6.5$ m, $h = 2.5$ m and $t = 1.25$ m. Substitute these values into expression 10 and approXimate to give expression 12.

Finally, to produce a table of values of α and the corresponding value of the lateral displacement of the point A, Author and then approXimate the expression

VECTOR ([180a/π, #12], a, 5 DEG, 45 DEG, 5 DEG) .

Interpretation

The results obtained above show that the lateral movement of the rear left corner of the vehicle can be quite considerable, especially if the vehicle moves off with the front wheels at an acute angle to the initial direction in which it is pointing; although it has to be said that in normal circumstances this is uncommon. Nevertheless, it is evident that it is unwise to get into a traffic situation in which you find yourself very close to the rear quarter of a large vehicle. Notice, too, that the results are independent of the turning speed (see expression 4). Do you think this is reasonable?

1: $(w + h)\,SIN(b) + t\,COS(b) - t$

2: $$\frac{SIN(a + b)}{SIN(a)}$$

3: $$\int_0^b \frac{SIN(a + b)}{SIN(a)}\,db$$

4: $$(w + h)\,SIN(b) + t\,COS(b) - t - w\int_0^b \frac{SIN(a + b)}{SIN(a)}\,db$$

5: $$\frac{COS(a)\,(w\,COS(b) - w)}{SIN(a)} + t\,COS(b) + h\,SIN(b) - t$$

6: $$\frac{d}{db}\left[\frac{COS(a)\,(w\,COS(b) - w)}{SIN(a)} + t\,COS(b) + h\,SIN(b) - t\right]$$

7: $- w\,COT(a)\,SIN(b) + h\,COS(b) - t\,SIN(b)$

8: $$b = ATAN\left[\frac{h\,SIN(a)}{w\,COS(a) + t\,SIN(a)}\right]$$

9: $$\frac{COS(a)\left[w\,COS\left[ATAN\left[\frac{h\,SIN(a)}{w\,COS(a) + t\,SIN(a)}\right]\right] - w\right]}{SIN(a)} + t\,COS\left[ATAN\left[\frac{h\,SIN}{w\,COS(a)\,+}\right.\right.$$

10: $$-\frac{\sqrt{\left[\frac{(t^2 - w^2)\,COS(a)^2 - 2\,t\,w\,SIN(a)\,COS(a) - t^2}{(h^2 + t^2 - w^2)\,COS(a)^2 - 2\,t\,w\,SIN(a)\,COS(a) - h^2 - t^2}\right]}\,((h^2 + t^2 -}{SIN(a)\,(w\,COS(a) + t\,SIN(a))}$$

11: $$-\frac{\sqrt{\left[\frac{(1.25^2 - 6.5^2)\,COS(a)^2 - 2\;1.25\;6.5\,SIN(a)\,COS(a) - 1.25^2}{(2.5^2 + 1.25^2 - 6.5^2)\,COS(a)^2 - 2\;1.25\;6.5\,SIN(a)\,COS(a) - 2.5^2 - 1.2}\right.}}{SIN(a)\,(6.5\,C}$$

12: $$\frac{\sqrt{\left[\frac{651\,COS(a)^2 + 260\,SIN(a)\,COS(a) + 25}{551\,COS(a)^2 + 260\,SIN(a)\,COS(a) + 125}\right]}\,(551\,COS(a)^2 + 260\,SIN(a)\,COS(a}{4\,SIN(a)\,(26\,COS(a) + 5\,SIN(a))}$$

COMMAND: **Author** Build Calculus Declare Expand Factor Help Jump soLve Manage
Options Plot Quit Remove Simplify Transfer moVe Window approX
Compute time: 1.3 seconds
Simp(11) Free:94% Derive Algebra

Figure 2.4 DERIVE's optimization of the left shift of A

LIVERPOOL JOHN MOORES UNIVERSITY
LEARNING SERVICES

Further Work

1. Investigate the solution for large vehicles of other dimensions.
2. Investigate the solution of the problem for the alternative turning strategy suggested. (The authors investigated the strategy α = constant as it is the more realistic of the two.) Do the two strategies lead to a significant difference in the amount of corner movement predicted ?

2.3 GOOD SHOT, SIR!

Introduction

Games such as snooker and billiards provide a rich source of both trigonometric problems and interesting applications of mechanics. Here we shall concentrate on their trigonometric potential.

This problem is about investigating the degree of precision required to execute snooker shots successfully from different points of the table. For the purpose of discussion we shall consider shots which are attempted into a corner pocket. You could adapt the analysis for shots into the side pockets (which are of a different size from the corner pockets). It is apparent from Fig. 2.5(a) that the degree of difficulty of the shot (and thus the level of precision required) varies from place to place (consider the points L, M and N in Fig. 2.5(a) – from which point would you find it easiest to play a successful shot?).

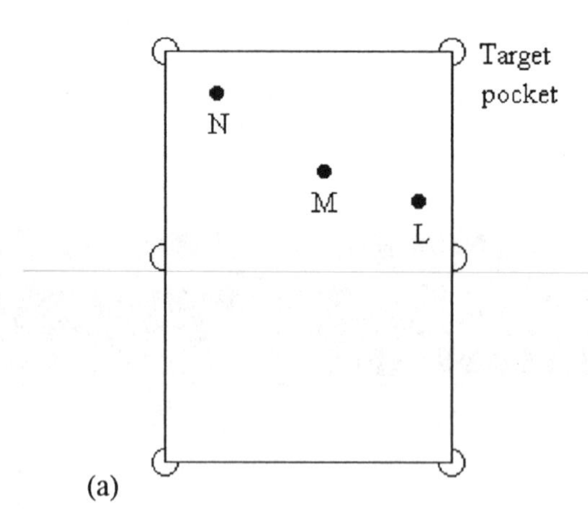

Figure 2.5 Snooker shots into a corner pocket

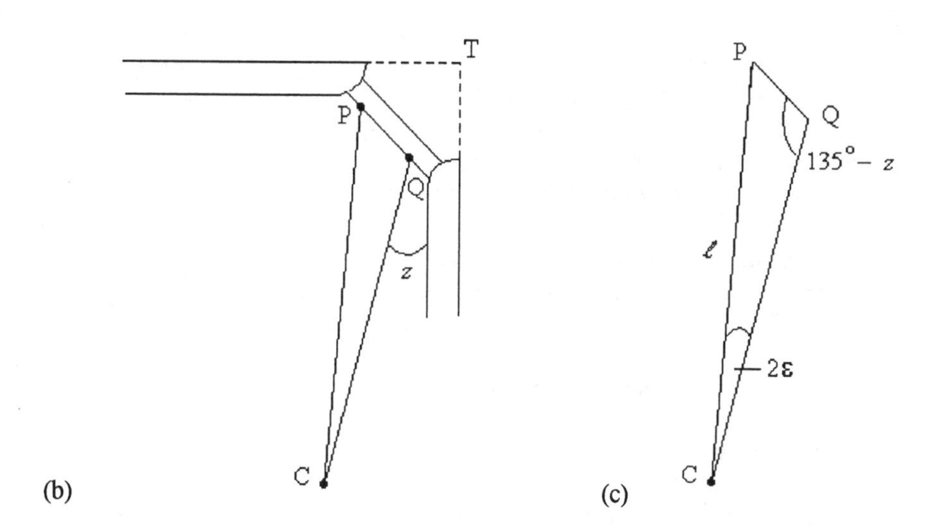

Figure 2.5 Snooker shots into a corner pocket (continued)

The Problem

The task is to develop a model which can be used to predict the precision necessary to successfully 'pot' a ball which is played from any point C on the table.

Setting up a Model

The width across the jaws of a corner pocket is only slightly greater than the diameter of the ball so the level of precision required must be high. The features on which the precision is likely to depend are as follows:

- the distance of the ball from the corner of the table (ℓ inches),
- the angle between a line drawn from the ball to the pocket and a side of the table (z degrees),
- the width across the jaws of the pocket,
- the diameter of the ball.

The last two features are in fact defined by the rules of the game and are, respectively, 3.5 inches and 2.0625 inches. Since inches are the units quoted in the rules it seems pointless to convert them into SI units merely for the sake of convention. We therefore assume that the relevant variables are represented by the first two features, namely ℓ (inches) and z (degrees).

With the relevant features identified we next consider our modelling assumptions. We shall assume that the ball:

- is struck directly on a line joining its centre of mass to the corner pocket; if delivered by a cue then the blow is assumed to be delivered horizontally and in line with the centre of mass, thus avoiding the problem of spin,
- is struck hard enough to reach the corner pocket,
- does not bounce off the sides of the pocket as it enters,
- is far enough from the pocket for this distance to be considered large compared with the width across the jaws,
- travels along a straight line path to the pocket from the point at which it was struck and is not affected by the nap of the table.

We are now ready to move on to the model formulation stage.

Formulate a Mathematical Model

Consider, first, the entry of the ball to the pocket. The points P and Q on Figure 2.5(b) represent the extreme positions of the centre of the ball for a successful shot (they correspond to the ball just touching either jaw).

Since the ball is assumed to be a reasonable distance from the corner pocket (if not then the shot should not present a problem), then the distance CP will be approximately equal to the distance ℓ from the ball to the corner of the table. Applying the sine rule in triangle CPQ then gives

$$\frac{\ell}{\sin(135° - z)} = \frac{PQ}{\sin 2\varepsilon}$$

In a practical situation it is reasonable to assume that ε is small and hence, using the small angle approximation and the data introduced earlier, we can deduce that

$$\varepsilon = \frac{23 \sin(135° - z)}{32\ell} \qquad (2.4)$$

Use DERIVE's facilities and a range of values of ℓ and z to investigate the level of precision required to pot the ball successfully from different parts of the table.

Interpretation and Validation

As you would probably expect, in terms of z the shot requires least precision in the neighbourhood of the line $z = 45°$, while in terms of ℓ it requires less precision when the ball is closer to the pocket. Nevertheless in all cases the value of ε is small, indicating that the action is in fact highly skilled – it is certainly beyond the authors' capability to replicate a successful shot regularly.

For values of z between approximately 30° and 60° the triangle CPQ may be considered to be isosceles, which means that the player has an equal amount of latitude either side of the line CT. At the other extreme when the ball is close to either of the cushions adjacent to the corner pocket, the triangle CPQ is most definitely not even approximately isosceles and so the amount of latitude on one side of the line CT will far exceed that on the other side.

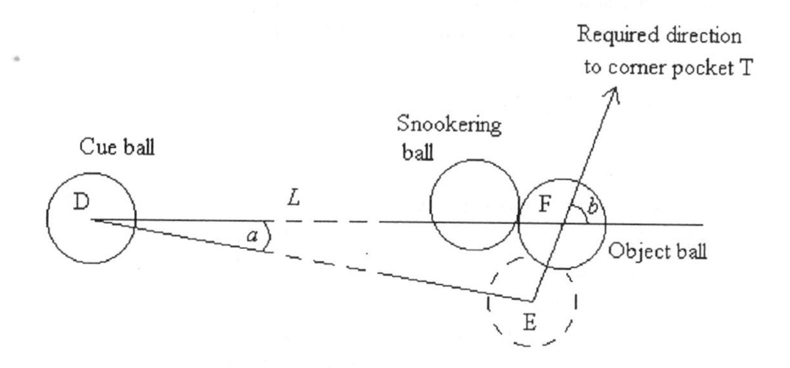

Figure 2.6 A strategy for dealing with a snookering ball

Revised Model

Fig. 2.6 shows a possible situation in the later stages of a match in which the cue ball does not have a clear path to the object ball. Because a certain amount of latitude is permitted in the movement from F to T (investigated in our initial model) there must be a consequent latitude in the angle a which determines the direction in which the cue ball is struck to impact with the object ball at position E. Investigation of this variation in a is left as a structured DERIVE activity for the reader as follows:

(i) By applying the sine rule to triangle DEF of Figure 2.6 show that

$$\frac{d}{\sin a} = \frac{L}{\sin(a+b)} \tag{2.5}$$

where d represents the diameter of the ball and L denotes the distance DF.

(ii) Use DERIVE's soLve facility to solve equation (2.5) for a to obtain the result

$$a = \operatorname{atan}\left(\frac{d \sin b}{L - d \cos b}\right) \tag{2.6}$$

followed by the Calculus facility to obtain an expression for $\mathrm{d}a/\mathrm{d}b$.

(iii) Recall that small changes in the quantities a and b, denoted by Δa and Δb respectively, are related to the derivative da/db by the approximation

$$\Delta a \simeq \frac{\mathrm{d}a}{\mathrm{d}b} \cdot \Delta b$$

and hence show that the variation in a consequent upon the variation Δb in the required direction of the shot is given by

$$\Delta a \simeq \frac{\Delta b \cdot d(d - L\cos b)}{2dL\cos b - d^2 - L^2} \tag{2.7}$$

The expression for Δb is given by 2ε, where ε is given in equation (2.4) and a is given in terms of b, L and d by equation (2.6).

(iv) Investigate the variation in a for shots from different parts of the table by varying the values of b, d, ℓ, L and the angle z.

You could use the Manage, Substitute commands to substitute for d, ℓ, L and Δb into the expression for Δa and then Author and approXimate the expression

```
VECTOR(VECTOR(#n, b, π/6, π/4, π/36), z, π/12, π/4, π/36)
```

in which expression n contains the substituted expression for Δb, to produce a table of values of Δb for b and z values in the ranges [$\pi/6$, $\pi/4$] and [$\pi/12$, $\pi/4$] respectively.

Interpretation

Your results should again indicate that very high levels of precision are also required in the value of a. It is interesting to note that, for publicity purposes, robot manufacturers have attempted to program robots to perform the shot analyzed here and the robot is not always successful, such is the precision required!

The analyses presented here have been purely in terms of the geometry of the shots attempted. No account has been taken of the inertia of the ball, the friction force between the ball and the table, the nap of the cloth or any spin effects imparted to the ball(s). These features offer the opportunity for a wealth of interesting mechanics problems and the interested reader is referred to two excellent papers by Mackie, see references [1] and [2].

2.4 ROUNDABOUT MODELLING

Introduction

There is a wealth of interesting modelling and mechanics to be found within amusement parks, not least from the roundabout.

Today, many amusement parks contain sophisticated variations of the roundabout for which the following description is typical. The roundabout consists of a central hub O, to which several radial arms are attached. At the end of each of these radial arms three or four secondary arms extend outward, each one carrying a car at its outer end in which one or two passengers can sit, see Fig. 2.7. The ride experienced by a passenger is thus a combination of two horizontal circular motions, that of the radial arm with the circular motion of the secondary arm superimposed on it.

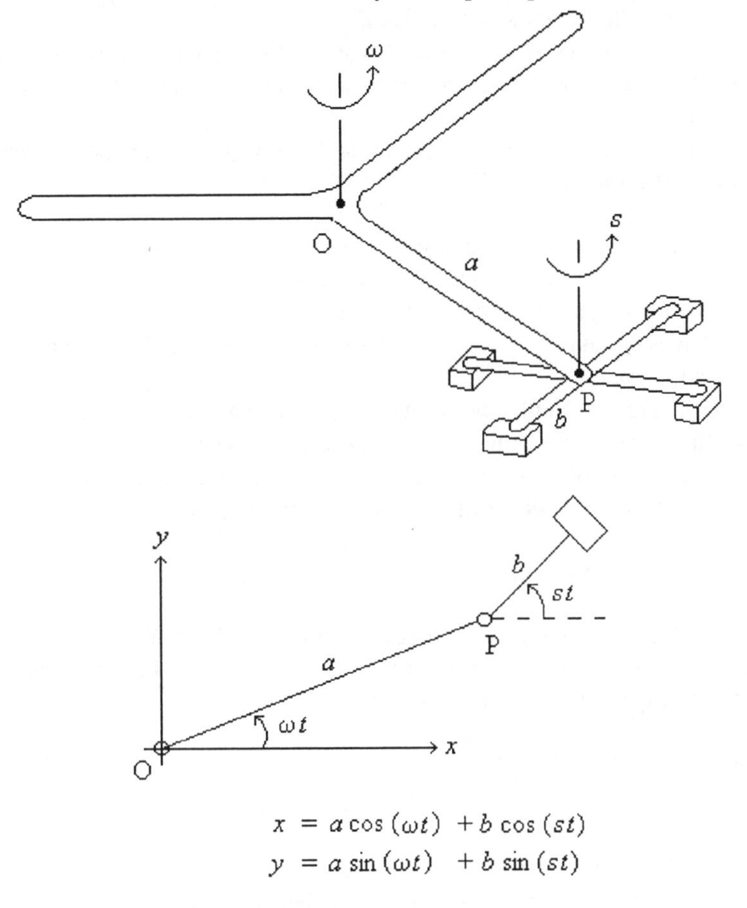

Figure 2.7 Schematic diagram of a roundabout and its associated trigonometry

Several questions immediately suggest themselves:

- As the radial arm rotates does the car loop or spiral?
- At which points of the car's path are its speed and acceleration greatest?
- What combinations of arm lengths and speeds of rotation will produce memorable rides?

Setting up a Model

First we need a feature list. The following features are likely to influence the ride experienced by the passengers:

- the length of the radial arm (a, metres) ,
- the length of the secondary arm (b, m) ,
- the angular speed and sense of rotation of the radial arm (w, rad s^{-1}) ,
- the angular speed and sense of rotation of the secondary arm (s, rad s^{-1}).

With the modelling features established we can now turn our attention to establishing our modelling assumptions:

- Since we are only considering the geometry of the ride we shall assume that there is no friction at the pivots O and P.
- Air resistance effects can be ignored.
- The position of a car will be measured relative to the fixed pivotal end of the radial arm.
- For simplicity, only one car will be considered in the analysis.
- Initially both the arms lie in a straight line from O to the car.
- Once the roundabout starts, the arms instantly rotate at their specified speeds; there is assumed to be no acceleration phase.

Formulate a Mathematical Model

Choose a set of Cartesian axes with origin at O and with the X axis aligned with the initial position of the arms. At some time t (seconds) after the roundabout starts to rotate, the coordinates of the car are thus given by

$$\begin{aligned} x &= a\cos(wt) + b\cos(st) \\ y &= a\sin(wt) + b\sin(st) \end{aligned} \qquad (2.8)$$

The velocity components of the car are found by differentiating equations (2.8) with respect to time, and the speed of the car is then given by $v(\mathrm{m\,s^{-1}})$, where

$$v = \sqrt{\left(\left(\frac{dx}{dt}\right)^2 + \left(\frac{dy}{dt}\right)^2\right)} \tag{2.9}$$

The central acceleration experienced by the passengers in the car is given by v^2/r (m s^{-2}) where r (m) is the distance from the car to O, the centre of rotation, and is derived from

$$r = \sqrt{(x^2 + y^2)} \tag{2.10}$$

Solution of the Mathematical Model

In order to give an idea of the diversity of paths available for the car we could investigate cases such as $a = b$, $a > b$, $a < b$, $w = s$, $w > s$, $w < s$ and with the senses of rotation of both arms either similarly directed or oppositely directed.

Figs. 2.8–2.11 show some possibilities and the reader is encouraged to investigate other cases. All that is needed is to Author equations (2.8) as a pair of parametric equations and Plot them.

The authors have substituted the value $a = 1$ into equation (2.8) before entering them into DERIVE in order to avoid difficulties with confusion with the inverse cosine. If you wish to preserve the generality, then you will need to Author the expressions as a* cos (wt), etc.

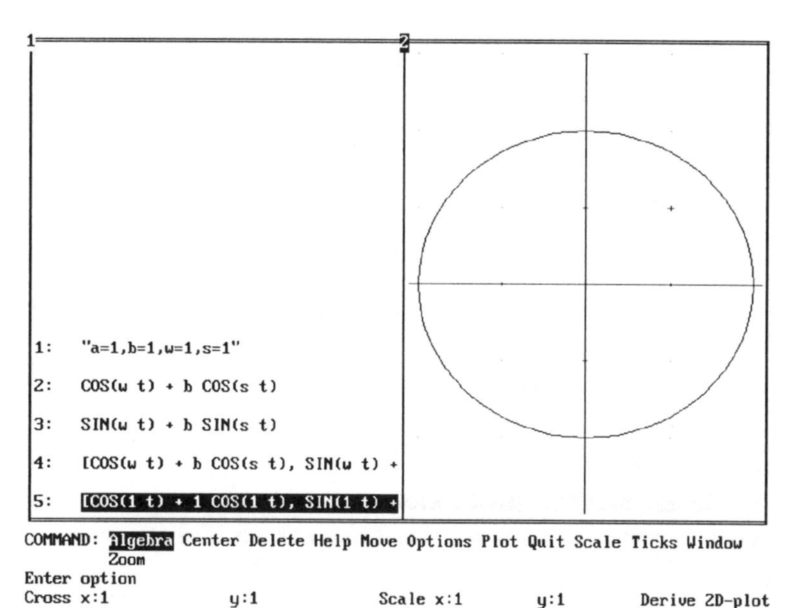

Figure 2.8 Path of car for equal length arms and equal angular velocities

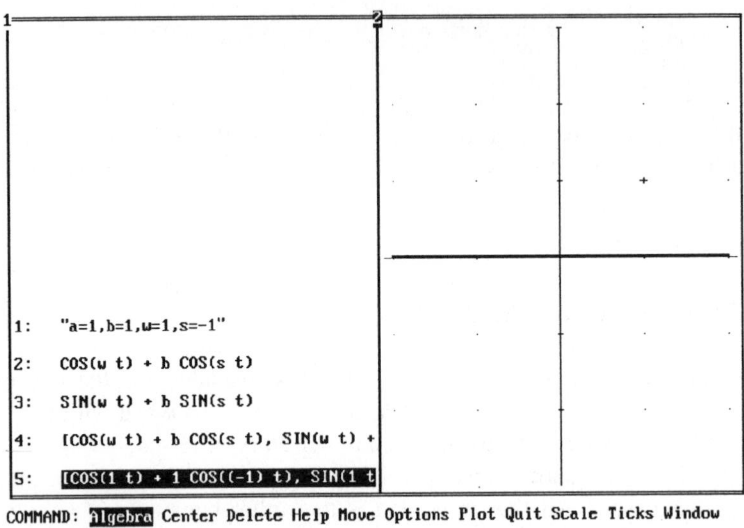

Figure 2.9 Path of car for equal length arms and equal but opposite angular velocities

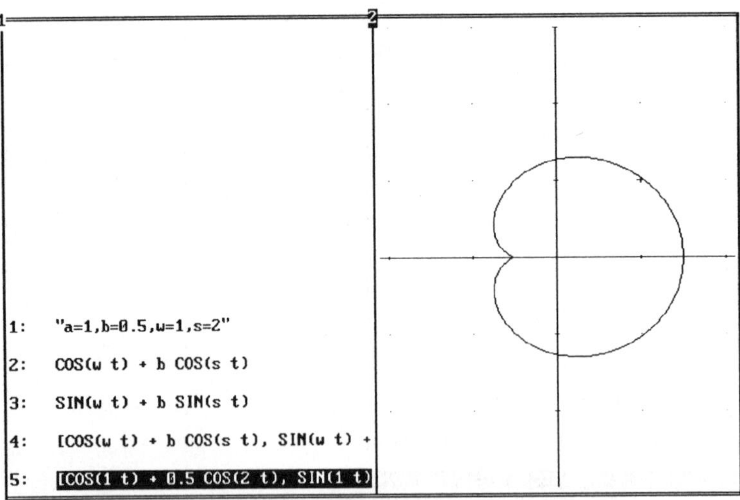

Figure 2.10 Path of car for unequal length arms and unequal angular velocities

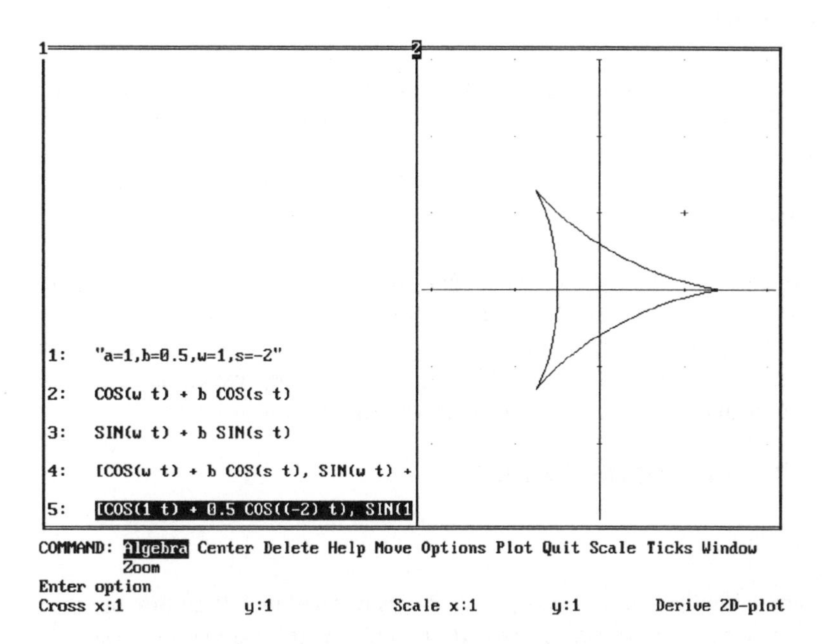

Figure 2.11 Path of car for unequal length arms and unequal and opposite angular velocities

Fig. 2.8 represents an uninteresting ride – it corresponds to equal length arms and equal angular velocities. Fig. 2.9 is even worse, just a straight line path (equal length arms and equal, but opposite, angular velocities). Of course, this case could not be realized in practice because the pivot O is in the way! Don't forget to check your solutions for this sort of reality! Things become a bit more interesting in Fig. 2.10, with unequal length arms and different angular velocities. Here the car spirals in towards O and then out again to pass through the initial position later and begin the ride again. Fig. 2.11 represents the case of both unequal length arms and oppositely directed angular velocities, and the path is much more dramatic. The car initially moves off right across the ride, comes to a stop and experiences a sudden change of direction, the effect being repeated three times per circuit – much more exhilarating!

For any one of these situations the speed could be investigated by Authoring and Plotting expression (2.9). The central acceleration could be similarly investigated by using expressions (2.9) and (2.10) to Author and then Plot the expression v^2/r.

The cross-wire could then be driven onto any of the extrema of the graphs and the corresponding value of t noted. Using Manage, Substitute this value of t could then be substituted into the parametric equations (2.8) to obtain the co-ordinates of the point corresponding to these extreme values of speed or acceleration.

Revised Model

Our model was developed in terms of the car which was initially in line with the pivot. If you wish to investigate the other cars this can be achieved by a simple time shift in the formulae for x and y.

Interpretation

It is evident from Figs. 2.8 – 2.11 that by judicious choice of the values of the parameters a, b, w and s the designer can produce a ride which will give the passenger any desired speed or acceleration experience. In terms of the magnitude of the acceleration, the results are not as high as the g loadings experienced on some of the other rides available within amusement parks–perhaps it is just as well!

2.5 TUMBLE DRIER DESIGN

Introduction

As the name implies, clothes in a tumble drier are dried by falling through the warm air within the drum. The clothes spend part of their time travelling on the wall of the drum and the remainder tumbling through the warm air.

If the speed of rotation of the drum is too slow, then the clothes will fall from the surface of the drum too soon and will not spend sufficient time in the warm air. On the other hand, if the speed of rotation is too great, then the clothes will adhere to the surface of the drum all the way round and the drum will then be acting as a spin drier.

The optimal speed of rotation of a tumble drier must therefore be somewhere between these two extremes. For a given radius , what is the optimal speed of rotation of the drum?

Setting up a Model

As usual, we begin by thinking about the features likely to influence the problem. Our thoughts produced the following list of features:

- the mass of the clothes in the tumble drier (m, kg),
- the radius of the drum (a, m),
- the angular speed of rotation of the drum (w, rad s^{-1}),
- the point at which the clothes leave the surface of the drum.

After a moment's thought, one realizes that the last feature above is related to w. We left it in our list to demonstrate that modelling is not always as orderly as a mathematical proof might be, and hindsight often reveals that some of the listed features are in fact dependent on each other.

With the relevant features established we need to formulate some modelling assumptions from which we shall develop the mathematical model. We assume the following:

- The clothes are modelled as a particle of mass m (kg) located on the inner surface of the horizontal drum. This avoids difficulties with the substantial size of the mass of clothes.
- The drum rotates at a constant angular speed of w (rad s^{-1}).
- Difficulties with tangling, which the authors have observed in some designs, can be overcome to some extent by periodically reversing the sense of rotation of the drum – a feature included in some machines, although it is ignored in this model.
- The clothes are partially 'lifted' by a series of radially directed small ledges positioned around the circumference of the drum.
- While in contact with the drum, the clothes do not slip and they rotate with the same angular speed as the drum. This is assisted to some extent by the ledges mentioned above.
- Once the clothes leave the drum surface they fall as a projectile until they encounter the drum surface again. Air resistance effects are ignored during this projectile phase.

Formulation and Solution of the Mathematical Model

The point at which the clothes leave the drum is determined by the angular speed of rotation of the drum. Subsequently, the clothes experience a drying effect throughout the flight time of the projectile phase. This drying effect is assumed to be maximized by maximizing the flight time.

A model will be developed to obtain an expression for the flight time as a function of w and then to maximize it with respect to w. The model will embrace several areas of mathematics:

- circular motion, to determine at what point the clothes leave the drum,
- projectile motion, to determine the subsequent path of the clothes,
- co-ordinate geometry, to establish when the clothes meet the drum again,
- calculus, to perform the maximization of the flight time.

The model and its solution will be developed as a series of stages and DERIVE activities for completion by the reader.

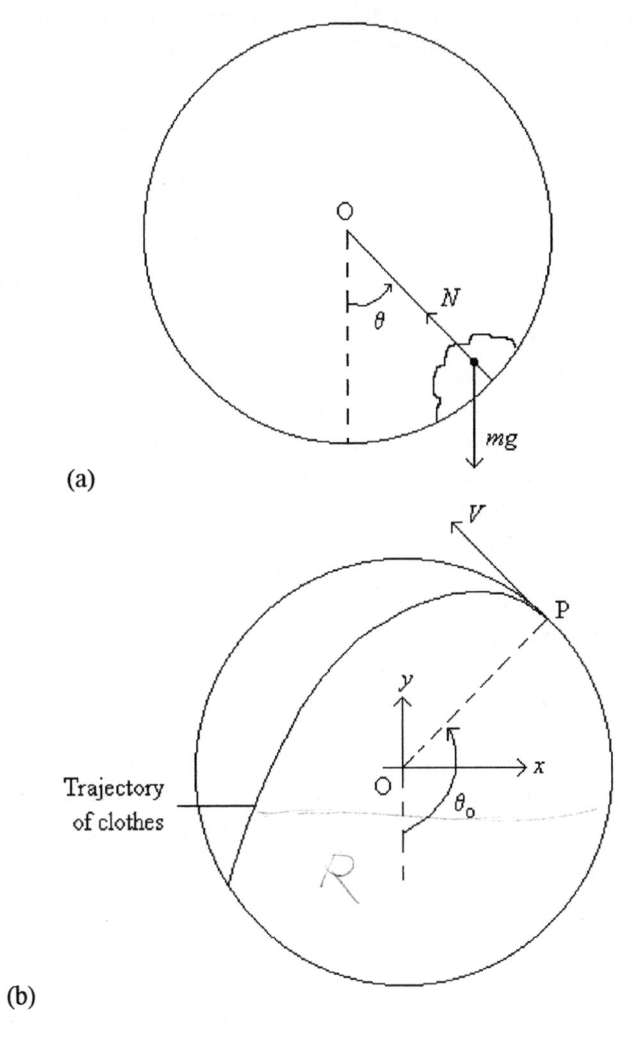

Figure 2.12 The tumble drier

Fig. 2.12(a) shows the clothes during the circular motion phase. Let N (Newtons) be the normal reaction of the drum on the clothes and let the angular position of the clothes be measured by θ (radians), measured with respect to the downward vertical through O, the centre of the drum.

(i) If the velocity of the clothes when in the position θ is V (m s^{-1}) show that

$$N - mg\cos\theta = mV^2/a$$

and deduce that the clothes leave the surface of the drum when at the angular position θ_0, where

$$\cos \theta_0 = -V^2/ag$$

(ii) With respect to a set of Cartesian axes OXY, show that the co-ordinates of the point P at which the clothes become a projectile are

$$(a \sin \theta_0 \,, \; - \; a \cos \theta_0)$$

and that at time t after the projectile phase has commenced, the co-ordinates of the clothes mass are given by

$$\begin{aligned} x &= a \sin \theta_0 + (V \cos \theta_0)t \\ y &= -a \cos \theta_0 + (V \sin \theta_0)t - \tfrac{1}{2} g t^2 \end{aligned} \qquad (2.11)$$

This completes the mechanics phase of the model and we next consider the geometry of the situation.

(iii) Explain why the clothes do not re-encounter the drum until

$$x^2 + y^2 = a^2 \qquad (2.12)$$

Using DERIVE, Author equation (2.12) and then use Manage and Substitute to replace x and y by equations (2.11). Simplify the result to show that the flight time t satisfies the quartic equation

$$\frac{1}{4} g^2 t^4 - g V \sin \theta_0 t^3 + t^2 (V^2 + ag \cos \theta_0) = 0$$

Use the result of (i) to further simplify the above equation and establish that the flight time is given by

$$t = \frac{4 V \sin \theta_0}{g} \qquad (2.13)$$

(iv) Use Manage and Substitute to replace V and $\sin \theta_0$ in favour of expressions involving the angular speed w to show that

$$t = \frac{4aw}{g} \sqrt{\left(1 - \frac{a^2 w^4}{g^2}\right)} \qquad (2.14)$$

(v) It just remains to maximize t with respect to w and this can be achieved using DERIVE's Calculus, Differentiate and soLve facilities.

But wait, don't abdicate all thought just because a package is being used. Maximizing t is equivalent to maximizing t^2 which would not be as complicated an expression as the square root is removed (see equation (2.14)). Therefore, it would be simpler to apply DERIVE's facilities to the expression

$$t^2 = \frac{16\,a^2 w^2}{g^2}\left(1 - \frac{a^2 w^4}{g^2}\right)$$

Show that this results in a maximum value of t when

$$w = (g^2/3a^2)^{1/4}$$

it being left as another exercise to confirm that it is a maximum value which is produced!

Interpretation and Validation

The flight time of the clothes through the warm air within the drum which, it is assumed, is equivalent to the drying time is maximized if the speed of rotation of the drum is $(g^2/3a^2)^{1/4}$ rad s^{-1}.

Domestic tumble driers have a drum radius of approximately 0.2 m. Our formula for w then gives a speed of rotation of 5.4 rad s^{-1} or, equivalently, about 52 rpm. This compares well with the figure of 58 rpm measured on the tumble drier belonging to one of the authors.

2.6 PURSUIT CURVES

Introduction

Today, the use of ground-to-air and air-to-air missiles to shoot down hostile aircraft is commonplace. The missile's navigation system uses sophisticated electronics and computers to first locate the target, then 'lock on' to it and thereafter alter the missile's course continuously so that it always points directly at the aircraft, regardless of any evasive tactics employed by the pilot.

It is interesting to examine the shape of the missile's flight path as this is likely to be very different from the projectile trajectories encountered elsewhere in this text.

Such paths are known as pursuit curves and we shall investigate this class of problem by outlining the steps required to develop a model of an equivalent ornithological interception.

The Problem

A hawk sees a sparrow resting in a line of trees 50 m away. At the instant that the hawk decides to attack and sets off towards the sparrow, the sparrow flies away along a line at an angle α to the line of trees. Develop a model with which to investigate the possibility of interception of the sparrow by the hawk.

Setting up a Model

The following is a list of the features that we considered to be of relevance:

- the speed of the sparrow (u, m s^{-1}),
- the speed of the hawk (v, m s^{-1}),
- the frequency with which the hawk adjusts its flight direction,
- the initial separation of the two birds (50 m).

With the relevant features identified we next consider the modelling assumptions to be made in order to produce a manageable initial model. We shall assume the following:

- The hawk always flies directly at the sparrow and repeatedly adjusts its direction to achieve this.
- The sparrow does not take any evasive action.
- The flight path of each bird is in the same two-dimensional plane.

Clearly, these last two assumptions are unrealistic and have been made purely in the interests of simplicity of the model whose development we next outline.

Formulate a Mathematical Model

A preliminary graphical approach can often help in clarifying initial thoughts about a model. Fig. 2.13 shows the development of the hawk's flight path if the sparrow flies away at right angles to the line of trees and if the hawk changes its direction every second so that it is directed at the sparrow. Such a large time step means that it is in fact always directed at a point somewhat behind the sparrow and hence the model will need to be developed in terms of a much smaller time step, say δt seconds.

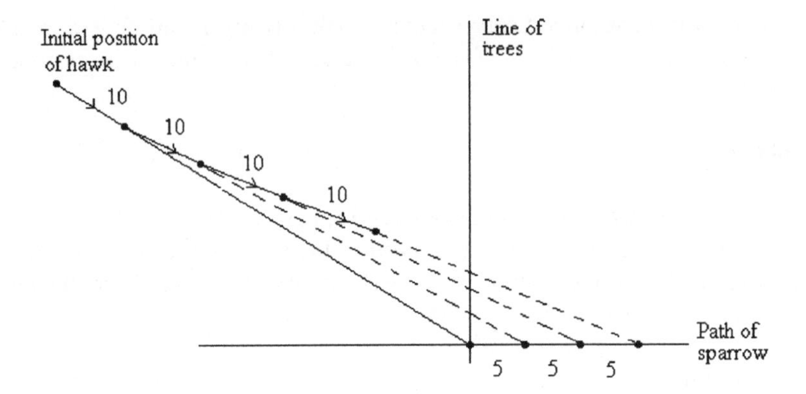

Figure 2.13 Adjustments to hawk's flight path made at one second intervals

The diagram has been drawn using a hawk speed of 10 m s^{-1} and a sparrow speed of 5 m s^{-1}. In developing our model it would be more general if the sparrow flew away in some constant direction inclined at an angle α to the line of trees. Fig. 2.14 shows the positions of the hawk and sparrow after the ith time interval. The positions are measured relative to a set of axes in which the origin is at the initial position of the sparrow and the y axis is along the line of trees. Relative to these axes the sparrow is at (x_i, y_i) and the hawk is at (X_i, Y_i).

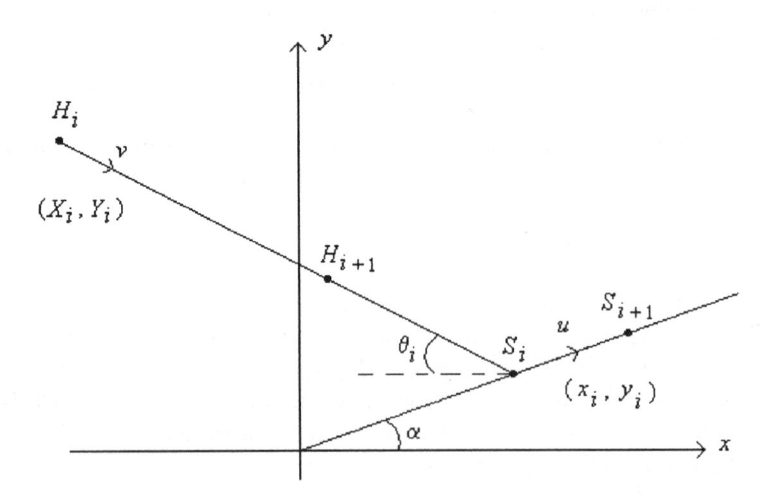

Figure 2.14 The geometry of the hawk–sparrow interception.

Show that the co-ordinates of the sparrow at S_{i+1} are

$$x_{i+1} = x_i + u\delta t \cos \alpha$$

$$y_{i+1} = y_i + u\delta t \sin \alpha \tag{2.15}$$

and that the co-ordinates of the hawk at H_{i+1} are

$$X_{i+1} = X_i + v\delta t \cos \theta_i$$

$$Y_{i+1} = Y_i - v\delta t \sin \theta_i \tag{2.16}$$

where

$$\theta_i = -\operatorname{atan}\left(\frac{y_i - Y_i}{x_i - X_i}\right) \tag{2.17}$$

Solution of the Mathematical Model

If the hawk is initially 50 m away from the sparrow in a direction inclined at 30° to the left of the line of trees, show that the initial values of the coordinates are

$$x_0 = 0 \text{ m}, \; y_0 = 0 \text{ m} \; ; \; X_0 = -25 \text{ m}, \; Y_0 = 43.3 \text{ m}$$

Use DERIVE's graphics capability to display the paths of the two birds on the same axes (for different speeds and time steps) in order to investigate the possible interceptions. To get you started, the velocities used in the initial graphical construction may be considered to be representative.

Interpretation

Your results should show the value to the hawk of frequent updates in its direction of flight. For some velocities you may find that the sparrow escapes.

As indicated earlier, two of our modelling assumptions are somewhat unrealistic. Firstly, the sparrow will obviously take some form of evasive action and, secondly, the real motion of the two birds will certainly be three-dimensional. Nevertheless, the model as presented does give considerable insight into both the geometry and dynamics of pursuit curves.

Further Work

Revising the model to produce a realistic three-dimensional one would prove extremely complicated – think of the complexity of missile guidance systems! However, you could postulate a two-dimensional evasion strategy for the sparrow and incorporate it into the analysis of Figure 2.14.

You could also investigate how the flight paths would be affected by a wind blowing across the line of trees at an angle β (a typical wind velocity is of the order of $4\ \text{m s}^{-1}$).

2.7 STAGGERING THE START

Introduction

Track races at athletics meetings always finish on a line drawn at right angles to the inner edge of the inside lane and extending across all the lanes of the track. Since athletes in 200 m and 400 m races remain in their own lane for the whole race they must each start from a different position in order that they have each travelled the event distance by the time they reach the finishing line. As we move outwards from the inside lane (lane 1) the distance ahead of the point where the runner in lane $n + 1$ starts relative to the runner in lane n is called the stagger.

The Problem

Develop a model with which to determine the stagger for each of the lanes of an athletics track (assumed to be a 400 m track) for the 200 m event.

Setting up a Model

Some features thought likely to be of relevance to this problem are

- the number of lanes,
- the width of each lane,
- the lengths of the straights,
- the radii of the ends (assumed circular),
- the path followed by an athlete within a lane, ...

There is perhaps one feature which is even more fundamental than any of the above and that is the answer to the question 'What exactly is meant by saying that a track is a 400 m track? Just where is the 400 m measured?' A reference such as [3] provides the answer to this question (0.3 m out from the kerb edge) and other information about the features listed above:

- Tracks usually have six or eight lanes.
- The recommended width of each lane is 1.22 m.

- The ends are semicircular with radii of between 32 m and 42 m for a 400 m circuit. The preferred radius is 36.5 m; anything less causes problems with 'running the bends'.
- The athlete in the inside lane is assumed to run on a line 0.3 m from the kerb edge (in practice this is not constant as, on the bends, the athlete will tend to run tangentially to the inner edge at some point).
- Athletes in all other lanes are assumed to run on a line 0.2 m from the inner edge of their lanes.
- The rules do not specify the length of the straight (nor is it necessary since the length of a circuit is known and the radii of the ends are specified).

If we assume that the track has six lanes, then the other assumptions required for the development of a model are in fact embedded within the data presented above.

Formulation and Solution of a Mathematical Model

We shall denote the various variables as follows:

- length of straight L (m) ,
- radii of ends R (m) ,
- lane number n ($n = 1,2, ..., 6$),
- radius of path in lane n R_n (m),
- width of lane W (m),
- amount of stagger in lane n S_n (m).

In one lap the runner in the inner lane (lane 1) travels the length of two straights plus two semicircular ends of radius (R+0.3) m. Thus

$$2L + 2\pi(R+0.3) = 400$$

Assuming that $R = 36.5$ m, the above equation implies that the length of the straight is 84.39 m. Tracks are constructed so that this distance is easily extended to provide straight lanes for the 100 m and 110 m hurdles events. To develop the model of the lane stagger further, we now need to specify the distance of the event (200 m, 400 m, ...) since each event involves the athletes in running around different numbers of bends (and in a manner which is in accordance with the regulations).

We begin our analysis by considering the 200 m event. This starts on the far side of the track, remote from the finishing straight. Assuming that the athletes actually follow paths as discussed in 'Setting up a Model' then the athlete in lane n ($n = 2,3, ...,6$) covers the finishing straight (L m) plus a semicircle of radius R_n (m) less the amount of stagger S_n (m) for that lane (see Fig. 2.15) so that

$$L + \pi R_n - S_n = 200$$

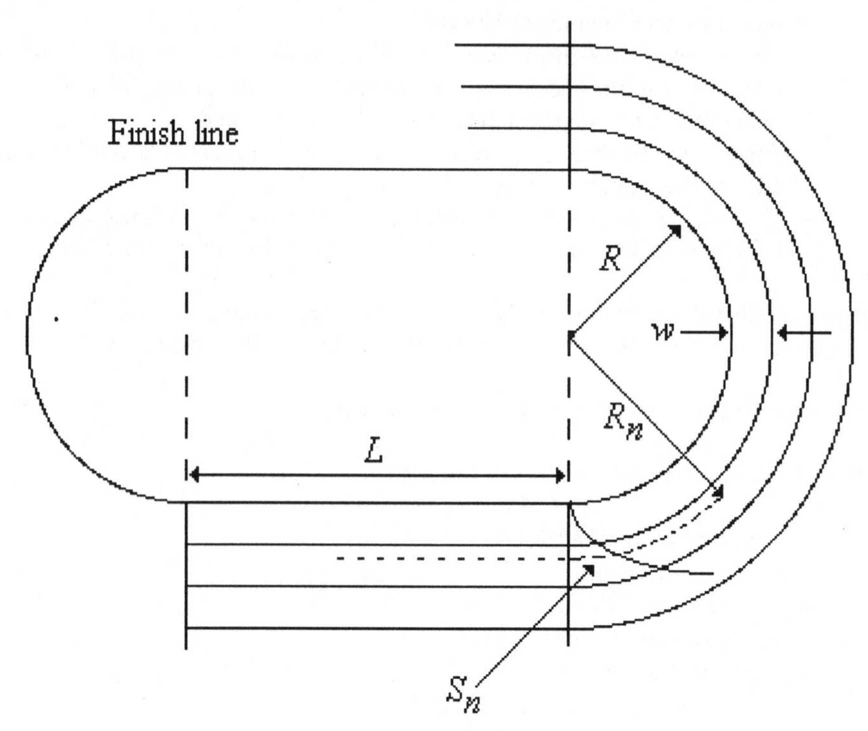

Figure 2.15 200 m stagger

where

$$R_n = R + W(n-1) + 0.2$$

and hence

$$S_n = \pi W(n-1) + L + \pi(R + 0.2) - 200\,, \quad n = 2,3, \ldots 6$$

It is left as an exercise for the reader to use DERIVE to generate a table of values of the amount of stagger required in each of lanes 2–6 for the track of dimensions obtained earlier (or for your local track if you know its values of L and R).

From the groundsman's point of view the values provided by S_n are inconvenient as they are measured not along the lane boundaries but along concentric arcs of radius 0.2 m greater than each lane boundary. Can you extend the model developed so far in order to provide the groundsman with corresponding staggers measured alongside the lane boundaries?

Interpretation and Validation

Your model can be validated by comparing the results with measurements made at your local track. Your visit should also reveal that there is a bewildering array of start lines marked on the track. What are they all for? For example, how much lane stagger is required for the 400 m event? This is left as a simple exercise for you to consider.

Extending the Model

In the 800 m event, the athletes have two laps in which to overcome the effects of starting in different lanes. The question arises as to how much lane stagger is therefore required at the start of this event.

Once again, it is useful to consult the rules book (equivalent to the modeller sometimes needing to consult an expert on the subject being modelled). The rules state that athletes can break from the start (i.e. they can all move across to the inner edge of the track immediately) except in the case of international competitions in which the first bend has to be run in lanes. We shall outline the solution to the 800 m stagger problem assuming that the athletes break from the start (the solution corresponding to an international competition is left for you to consider).

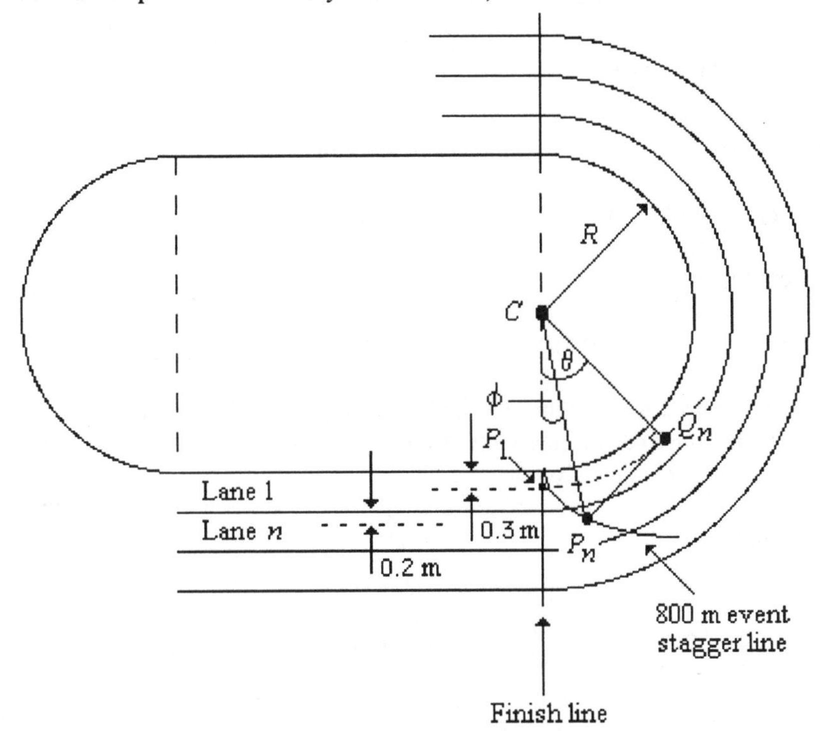

Figure 2.16 800 m stagger

Fig. 2.16 shows the start–finish area and the 800 m stagger line drawn across the lanes. The point P_1 represents the start position of the athlete in the inside lane, P_n represents the start position of the athlete in lane *n*. At the start of the race the athlete starting from P_n will move across the (n-1) lanes to his left to take up a position on an arc in the inside lane, assumed to be of radius $(R+0.3)$ m. His tangential direction P_nQ_n across to this inner path must therefore be such that

$$\text{The tangential distance } P_nQ_n = \text{the arc length } P_1Q_n \qquad (2.18)$$

so that both runners have covered the same distance by the time they reach Q_n – thereafter they both run the remainder of the race on a line 0.3 m out from the kerb of the inside lane.

From Figure 2.16

$$\text{Arc length } P_1Q_n = (R+0.3)\,\theta \qquad (2.19)$$

where θ is the angle subtended at C by the arc P_1Q_n.

From triangle CP_nQ_n Pythagoras' theorem gives

$$\text{Tangential distance} P_nQ_n = \sqrt{\left(R_n^{\ 2} - (R+0.3)^2\right)} \qquad (2.20)$$

where R_n is the radius of the path followed by the athlete in lane *n*, i.e.

$$R_n = R + W(n-1) + 0.2 \qquad n = 2,3,\ldots,6$$

Again from Figure 2.16,

$$\theta = \phi + \text{acos}\,((R+0.3)/R_n) \qquad (2.21)$$

The stagger required for the athlete P_n in lane *n* is given by $R_n\phi$, measured along a line 0.2 m from the inner edge of the lane. Combining equations (2.19), (2.20) and (2.21) the stagger in lane *n* is thus given by

$$stagger = R_n\,\phi$$

$$= \frac{R_n}{(R+0.3)}\sqrt{\left(R_n^{\ 2} - (R+0.3)^2\right)} - R_n \cdot \text{acos}((R+0.3)/R_n)$$

where

$$R_n = R + W(n-1) + 0.2\ , \quad n = 2,3,\ldots,6$$

Rather than immediately using DERIVE to generate a table of values of the stagger for each lane it would again be more useful to determine the corresponding distances measured along the lane boundaries. It is left as an exercise for the reader to determine these distances and produce a table of values. Once again, the results can be validated by a visit to your local track.

Further Work

There are plenty of other track events up to 10 000 m which require some measure of stagger and can be analyzed by developing the ideas outlined in this example.

Another variation is to consider staggers and lane markings for indoor track events where the number of lanes is fewer, the bends are banked and the track is usually a 200 m track.

References

[1] Mackie A.G., 'The mathematics of snooker', *IMA Bulletin*, **18**, 82–89 (1982).

[2] Mackie A.G., 'Trajectories of spheres moving on a rough surface', *IMA Bulletin,* **22**, 105–108 (1986).

[3] The Diagram Group, *Rules of the Game,* Collins Willow, London (1991).

3

Algebraic Models

3.1 INTRODUCTION

In this chapter we shall investigate modelling problems that require a knowledge of mathematical functions, vectors and matrices and their algebraic manipulation. Wherever possible, readers are encouraged to use the DERIVE graph-plotting facilities to plot proposed models and solutions in order to analyze the results both analytically and visually. The reader is also encouraged to attempt the DERIVE activities suggested after each case study.

3.2 FORMULA FUNDING OF SCHOOLS

The Problem

Many schools within the United Kingdom are under the control of a local education authority and receive an annual budget which must provide for school equipment, teachers' salaries, etc. The budget usually covers the financial year running from April to the following March, even though the academic year runs from September to the following August. The size of the budget is calculated using a formula designed to ensure careful budgetary planning, albeit from contrasting perspectives: schools would wish to maximize their budgets whereas a local authority would probably wish to ensure that its total education budget is spent efficiently.

The major factor in the funding formula is the number of pupils in the school. Each pupil is 'worth' a certain amount of money per annum, with this amount varying according to the pupil's age. The budget also contains a basic sum per annum which varies slightly with such things as inflation. Extra budgetary income can be obtained if the school is designated as being 'small', i.e. if the number of pupils is less than a set figure, N say. This extra income is given as a bonus for each pupil, but this reduces to zero as the number of pupils increases to N. There is also a bonus payable for pupils regarded as having special learning needs.

The aim of this case study is to develop a model incorporating the current funding formula which can be used to aid both the schools and the local education authorities with their conflicting economic strategies. The implications of this and other possible funding formulae can then be established.

Setting up a Model

The problem statement indicates some important features needed in any model, but we shall list the major features and likely assumptions here:

- The number of pupils in the school; if this changes every September, how is the change accommodated in a budget calculated from April to March?
- The designation of a school as 'small' or not; how many pupils can a small school take before it loses its 'small' status?
- The phrase in the problem statement 'the bonus per pupil reduces to zero as the number of pupils increases ...'.
- The size of the basic sum and the funding per pupil.
- The number of pupils with special learning needs.
- The age distribution of the pupils.

From a school perspective the number of pupils is related to the number of teachers via student–staff ratios. In general, more pupils need more teachers which leads to a greater salary demand on the budget.

From a local authority viewpoint, too many 'small' schools may be expensive to maintain and amalgamation of schools may need to be considered. In each case, the proposed model must be able to aid planning. In the first instance, the following assumptions will be made:

- The number of pupils in the school will be an average of its roll in two consecutive academic years and will apply for the whole of a financial year. Let x be this number of pupils.
- N pupils is the maximum size for a small school, i.e. if $x \leq N$, then the school is 'small'.
- The basic sum per annum, £B, is independent of x but may vary each year.
- The funding per pupil, £u, is an average figure per pupil which is independent of age distribution but is possibly a function of x.
- Additional funding per pupil with special learning needs is likely to be small in the majority of schools and is therefore ignored.

Formulate a Mathematical Model

The model developed here closely follows that proposed in [1], in which this problem is considered in more detail.

First, we need to propose a model for $u(x)$ and in particular we need to cater for $x \leq N$ and $x > N$ separately.

For $x \leq N$, the phrase 'the bonus per pupil reduces to zero as the number of pupils increases to N' can be most easily modelled by a linear relationship

$u(x) = ax + b$ such that:

$u = K_0$ (a positive constant) when $x = 0$

$u = K$ (a positive constant) when $x = N$, where $K < K_0$

For $x > N$ no 'small' school bonuses apply and hence u is now independent of x. With this formulation we have

$$u(x) = \begin{cases} K_0 + \left(\dfrac{K - K_0}{N} \right) x & , \text{ for } x \leq N \\ K & , \text{ for } x > N \end{cases} \tag{3.1}$$

The annual school budget, £y, is then given by

$$y = x\,u(x) + B$$

B and K are amounts of funding that occur regularly and it is convenient to simplify the equations by setting

$$q = \frac{y - B}{K}$$

so that the expression for the school budget can be written more concisely as

$$q(x) = \begin{cases} x\left[\alpha + \dfrac{1}{N}(1-\alpha)\,x\right] & , \quad x \leq N \\ x & , \quad x > N \end{cases} \tag{3.2}$$

where $\alpha = \dfrac{K_0}{K} > 1$

From a school management point of view, it is useful to be able to see how the budget changes as extra pupils are enrolled into the school. If the number of pupils is increased by δx, then the increase δq in the budget q is approximated by

$$\delta q \simeq \frac{dq}{dx}\,\delta x$$

From (3.2), $\dfrac{dq}{dx}$ is given by

$$\frac{dq}{dx} = \begin{cases} \alpha + \frac{2}{N}(1-\alpha)x & , \quad x \le N \\ 1 & , \quad x > N \end{cases} \tag{3.3}$$

Solution of the Mathematical Model

It is of interest at this point to see how these functions vary as x increases and we shall do this using the parameter values quoted in [1], namely $N = 750$ and $\alpha \simeq 1.53$.

$u(x)$ and $q(x)$ are piecewise continuous functions of x and can be represented as DERIVE expressions either by using the IF function or the STEP(x) function. The latter is used here but the reader is encouraged to try both functions.

STEP(z) takes the value 1 if the argument z is a positive number and 0 if z is negative (strictly speaking the case $z = 0$ is undefined – the use of IF can overcome this). Hence, to plot $u(x)/K$ using the values $\alpha = 1.53$ and $N = 750$, Author the expression

$$(1.53 - 0.53x/750) \text{ STEP } (750 - x) \quad + \quad (1 - \text{STEP}(750 - x)) \tag{3.4}$$

Select Plot and Overlay and adjust the Scale to be x : 200 , y : 0.5 .

Use the Centre facility to position the axes so that $x = 0$ appears on the left-hand side of the screen with the x-axis somewhere near the bottom of the screen and Plot expression (3.4), see Fig. 3.1.

Similarly, to plot $q(x)$, Author the expression

$$x\left((1.53 - 0.53x/750) \text{ STEP } (750 - x) \quad + \quad (1 - \text{STEP}(750 - x))\right)$$

(note that this is simply x times expression (3.4)).

Select Plot and Overlay and adjust the scale to x : 200, y : 400. Centre the graph as before to obtain a plot for $q(x)$ as in Fig. 3.2.

dq/dx can be plotted in either of two ways. Either Author an expression for (3.3) involving the STEP or IF functions as above and plot the resulting function: (this is left as an exercise for the reader).

Alternatively, assuming the expression for $q(x)$ is represented by the DERIVE expression m, select Calculus, Differentiate, #m and enter variable : x and order : 1 and Simplify. Then use Plot, Overlay as before with scale x : 200, y : 0.5 to obtain Fig. 3.3.

Using this latter approach, DERIVE produces the expression for dq/dx as

$$\frac{53(x - 375) \text{ SIGN } (x - 750)}{75000} + \frac{94875 - 53x}{75000}$$

which makes use of the function SIGN(z) which has the value 1 if the argument z is positive and -1 if z is negative. Readers should convince themselves that the two approaches give equivalent expressions for dq/dx.

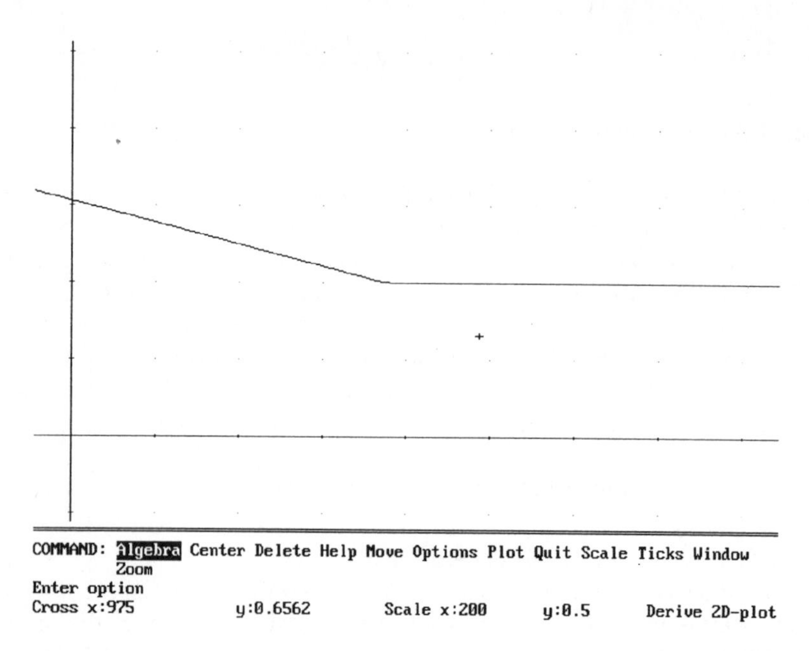

Figure 3.1 Funding per pupil as a function of the number of pupils

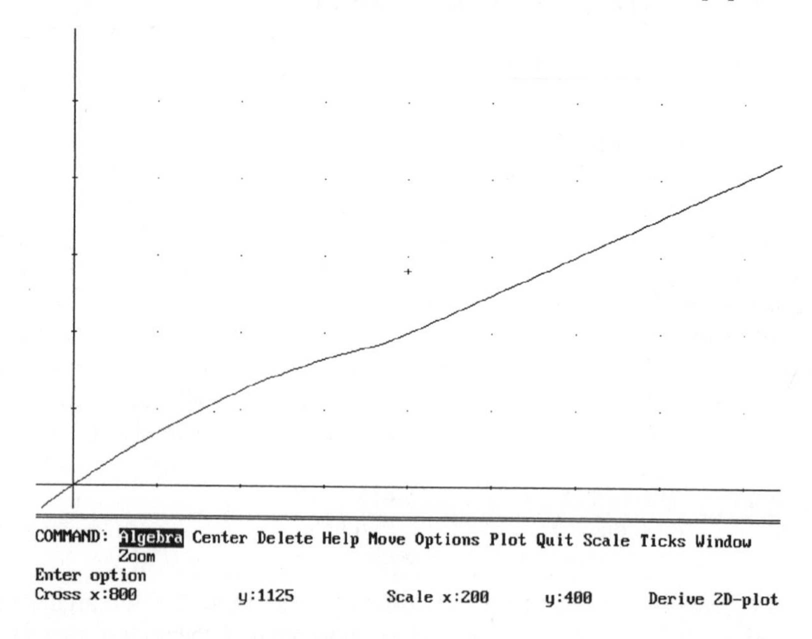

Figure 3.2 Normalized school budget as a function of the number of pupils

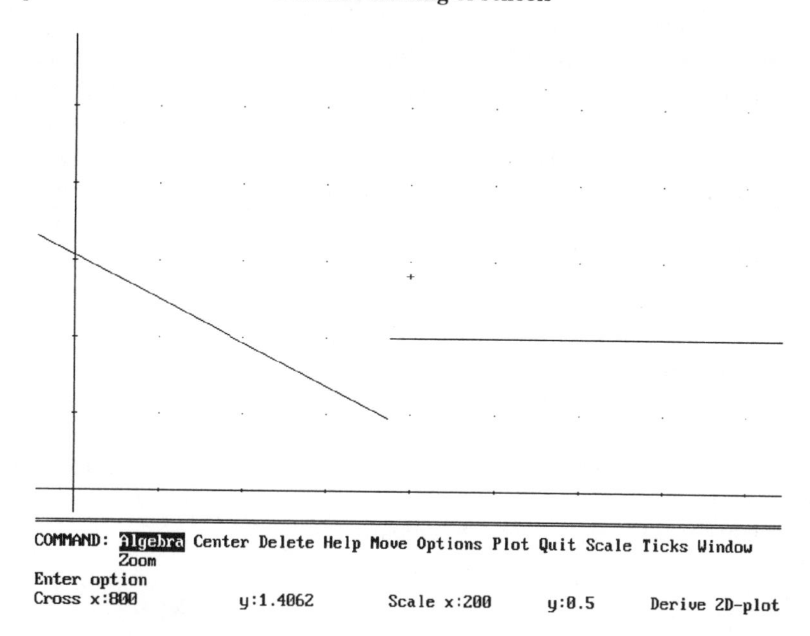

Figure 3.3 Rate of change of normalized school budget with respect to number of pupils

Interpretation

First let us try to answer the following question:

> As a School Manager of a small school with the average number of pupils very close to, but less than N (750 in our example), would you be keen to increase the number of pupils in the school?

An answer to this question can be obtained by inspecting the graphs of q and $\mathrm{d}q/\mathrm{d}x$. Note the 'bulge' in the graph of $q(x)$ for $x \leq N$ which is due to the effect of the small school status. By using the cross-hair on the DERIVE plot shown in Fig. 3.3, it can be seen that for x between $N/2$ and N, the increase in budget for each extra pupil (i.e. $\delta x = 1$ so that $\delta q \simeq \dfrac{\mathrm{d}q}{\mathrm{d}x}$), is less than it is if $x \leq N/2$ or $x \geq N$. In fact, for x very close to, but less than, N, the increase in budget per extra pupil is less than half the value for $x \geq N$. Thus the model suggests that the number of pupils should either be restricted to a figure 'comfortably' less than N or be increased to a figure 'comfortably' greater than N. (Perhaps the reader would like to interpret 'comfortably'!)

Of course, the question of school size is related to school costs via teachers' salaries, books, equipment, etc., and any increase in pupils is likely to increase such costs. Can the model be extended to incorporate such factors?

One way to proceed is as follows. Suppose that the amount spent on text books and equipment is a fixed amount £*p* per pupil, irrespective of the size of the school. Suppose that heating and lighting and other running costs are £*H*, irrespective of the school size. (You may argue that this is unreasonable in that more pupils implies more buildings, which implies more running costs – if you feel this way, incorporate this cost within the value of *p*.) Then, if these costs are deducted from the budget, the amount available for teachers' salaries, £*A*, is given by

$$A = \begin{cases} x\left[K_0 - p + \dfrac{(K-K_0)}{N}x\right] + B - H & , \quad x \le N \\ x(K-p) + B - H & , \quad x > N \end{cases}$$

The salary of each individual teacher depends upon his or her individual pay scale and his(her) position on that scale. If we assume that the average salary of a teacher is £*S*, then the number of teachers, *T*, able to be employed is

$$T = \frac{A}{S}$$

(To be precise, *T* equals the greatest integer less than or equal to *A*/*S* which can be represented in DERIVE as FLOOR (*A*, *S*).)

Schools usually talk in terms of the 'pupil–teacher ratio (PTR)', namely the number of pupils divided by the number of teachers. Thus the PTR is given by

$$PTR = \begin{cases} \dfrac{xS}{\left\{x\left[K_0 - p + \dfrac{(K-K_0)}{N}x\right] + B - H\right\}} & , \quad x \le N \\ \dfrac{xS}{\{x(K-p) + B - H\}} & , \quad x > N \end{cases}$$

Since in our assumptions $K_0 < K$, then the reader should be able to observe that this model suggests that small schools can afford to run at a lower PTR than large schools. (If not convinced, pick any positive values for S, K, K_0, B and H and plot *PTR* versus *x* using DERIVE.)

It should be evident from the interpretation of this model that the manager of a 'small school close to becoming large' can have serious difficulties with this funding formula. The difficulty is due to the discontinuities in some of the graphs, particularly $\dfrac{dq}{dx}$ versus x, near $x = N$.

Further Work

An obvious further consideration is to investigate whether it is possible to construct a model for the amount of funding per pupil $u(x)$ that will remove the problems noted above (and yet at the same time not increase the overall education bill for the local authority). The reader can pursue this via the following stages:

1. What functions, $u(x)$, can you think of that have the property of being a decreasing function of x (i.e. $\frac{du}{dx} < 0$) over the range $0 < x \leq N$? Write down some such functions.
2. One such possibility for $u(x)$ is

$$u(x) = a + b/x \quad , \quad a, b \text{ constants}$$

Determine conditions on a and b such that

$$u = K \text{ for } x = N \; ; \quad \frac{du}{dx} < 0 \quad \text{for} \quad 0 < x \leq N$$

3. You should find that $u(x)$ still contains an arbitrary constant in its definition for $x \leq N$. One way to fix its value is to equate the new $u(x)$ to that given by expression (3.1) for some value of x, say $x = 50$. Do this (using DERIVE to soLve equations if necessary) to obtain an expression for $u(x)$ in terms of x, K, K_0 and N as before.
4. Use DERIVE to Author and Plot your new expression for $u(x)/K$ using $N = 750$, $K_0/K = 1.53$ as before. Observe the shape of the graph and comment on whether

   (i) this model looks as if it will be less costly to the local authority 'on average' compared with the original model,

   (ii) the fact that $u(x)$ is infinite at $x = 0$ with this model can be accommodated in any way.

   If you feel the answer to (i) is NO, rework stage 3 equating expressions at some other value of x, $0 < x \leq N$, until your answer to (i) becomes YES.
5. With your new $u(x)$, re-calculate expressions for $q(x)$ and $\frac{dq}{dx}$ as before and plot them. Does this model remove the discontinuities found in the earlier graphs? Does this model ease the decision making process for schools with x less than but near to N?
6. Repeat stages 2–5 for another possible model

$$u(x) = a\,x + \frac{b}{x} \qquad \text{for } x \leq N$$

Does this offer any improvement over the two models already considered?

7. Rather than fix a constant within the model by equating to the original $u(x)$ for some x value, try instead to fix any arbitrary constant by choosing to satisfy the condition that dq/dx is continuous (i.e. has no breaks) for all values of x (including $x = N$).

   If possible, impose this condition on the models proposed for $u(x)$ in stages 2 and 5. For any new $u(x)$, re-examine the question posed in stage 4(i).

8. Repeat the whole process for any other $u(x)$ you may have proposed in stage 1. The challenge is to establish a function $u(x)$ that would be more satisfying to both schools and local authorities than those proposed by the authors!

3.3 PARALLEL ALGORITHMS AND TRANSPUTERS

The Problem

Readers may well be familiar with computers that contain a single processor and cater for SEQUENTIAL programming; all the tasks that make up a process or algorithm are performed one after the other with no opportunity for two or more tasks to be performed simultaneously.

In recent times, with the development of both computer hardware and software, other computing systems have been developed which contain a number of processors that can be linked together in a variety of ways with the capacity to communicate the data between the processors. Such a system allows for a number of tasks to be performed in PARALLEL and hence the possibility exists of improving the time taken to execute an algorithm. The savings in computation time may be significant, a bonus when results are needed within a time limit, as is the case, for example, in weather forecasting, image processing, etc.

Parallel processing does raise some interesting issues such as the following:

(i) Given a number of processors, how is the algorithm best 'split-up' among the processors in order to minimize the time to complete the algorithm?
(ii) How does the number of processors and their physical arrangement (i.e. the topology of the system) influence (i), if at all?
(iii) Is there any guarantee that the 'best' algorithm when programmed in a sequential manner will still be the best algorithm for parallel processing? (And if not, what is?)

In short, the performance of a proposed parallel algorithm is influenced by many factors and, like many areas of science, this technology would benefit from the generation of realistic models of performance in order to avoid excessive and expensive computational experiments.

As a specific illustration, we shall attempt to model algorithmic performance on a network of devices called transputers. A transputer is in effect a microcomputer on a chip, with each transputer having a processing capability (to perform the usual arithmetic operations) and a small amount of memory (to store data). The transputer has four links with two channels of data communication per link which can be used to connect to other transputers to build up any desired network topology. Fig. 3.4 illustrates the concept of a transputer, while Fig. 3.5 shows examples of typical topologies that can be constructed.

Factors of transputer technology that impact on any attempt to model algorithmic performance on a transputer network include the following:

- No transputer shares any memory with any other transputer.
- Data transmission between transputers is a bi-directional process along a channel.
- The transputer can transmit data and perform arithmetic in parallel.
- Data to be transmitted onto the transputer network are usually routed through a master or root transputer that is linked directly to a host computer to permit data input/output via the keyboard/screen.

Readers interested in transputer technology can obtain further information from references [2] and [3], for example.

Setting up a Model

Let us consider the computation of the arithmetic mean value, $\bar{x}$, of a set of numbers $x_1, x_2, \ldots, x_n$ using a network of transputers arranged in the form of a linear chain (see Fig. 3.5 (a)). The formula for computing this mean is given by

$$\bar{x} = \frac{1}{n}\sum_{i=1}^{n} x_i$$

What are likely to be important features in modelling this problem? The authors' list includes the following:

- The time taken to communicate a single data item – and does this vary with different data types, e.g. real numbers, integers etc., represented by different computer word lengths (16-bit, 32-bit, 64-bit representations, etc.)?
- The time taken to perform an arithmetic calculation – and does it vary for the different operations of addition, subtraction, multiplication, etc. (And with different data types)?

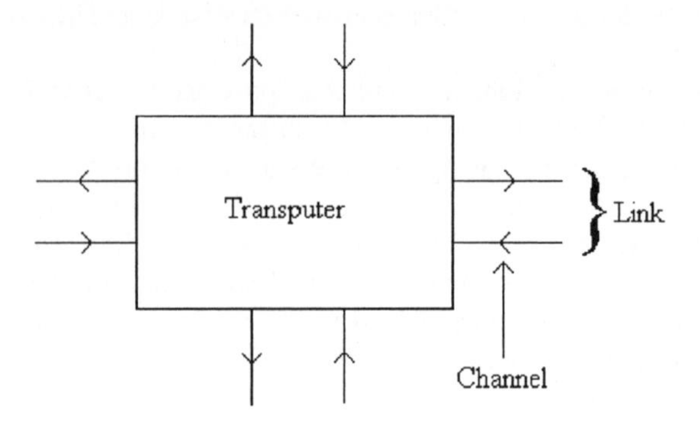

Figure 3.4 Conceptual diagram of a transputer

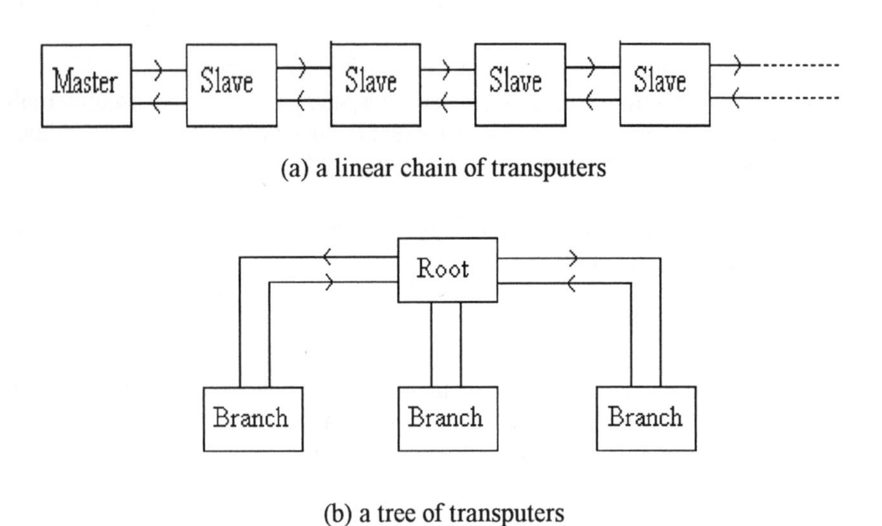

Figure 3.5 Examples of transputer topologies

- The storage of data on a transputer – how should the data be distributed among the various transputers in order that they can be optimally accessed at computation time?
- The declaration of data values – does it make any timing difference if values are declared as n single items or as elements $x[1]$, $x[2]$, ... ,$x[n]$ of an array $\boldsymbol{x}$?

- The synchronization process involved when data are communicated between transputers – is this influenced by the direction of data 'flow', parallel activity being undertaken by both the transmitting and receiving transputers, etc.?
- The computation strategy – how is the calculation to be performed on the transputer network?

A more detailed list could probably be generated by closer inspection of the transputer technology and answers to some of the above issues can be found in [4]. To make progress, the following assumptions will be made and notation used:

- The data are all real numbers with 32-bit representation.
- The time taken to transmit n such data values from one transputer to another varies linearly with n and is represented by

  $$t_n = t_s + n\, t_c$$

  where t_s (seconds) is a constant 'start-up' time and t_c (seconds) is the time to communicate a single data item once the transputers are ready to transmit and receive.
- The average time for an arithmetic operation between pairs of these data values is given by t_f (seconds). This average time will include any 'reading from' and 'writing to' memory, etc., and is assumed to be the same for all arithmetic operations.
- All transputers are of the same type and have the same timing characteristics.
- The linear chain of transputers consists of one 'master' and $(p-1)$ 'slaves', i.e. p transputers in all. Initially, all the data values reside on the master transputer.
- The computation strategy to compute $\overline{x}$ is as follows:

  Stage 1: The master processor (transputer) exports $(p-1)$ 'packets' of m values to each of the slaves so that on completion each of the p transputers has m values, where $mp = n$. It is further assumed that n is exactly divisible by p.

  Stage 2: In parallel, each transputer sums its own m data values.

  Stage 3: The results of each 'partial' sum are passed back to the master transputer.

  Stage 4: $\overline{x}$ is calculated on the master by summing these partial sums and dividing by n.

  Note: we are making use of the fact that

$$\frac{1}{n}\sum_{i=1}^{n} x_i = \frac{1}{n}\left\{ \sum_{i=1}^{m} x_i + \sum_{m+1}^{2m} x_i + \ldots \sum_{(p-1)m+1}^{n} x_i \right\}$$

In this case we can establish whether some of the above simplifying assumptions are reasonable by actually measuring the times to communicate n data items between two transputers and also to perform some arithmetic operations by using an accurate timer that forms part of the transputer hardware. Such experiments, detailed in reference [4], for a certain type of transputer, indicate that the communication and arithmetic operation models are reasonable, with appropriate values for the constants of:

$$t_s \simeq 12.2 \times 10^{-6}\,, \quad t_c \simeq 2.2 \times 10^{-6}\,, \quad t_f \simeq 2.64 \times 10^{-6}\ \text{s}$$

Hence we can proceed to estimate the performance of the proposed computational strategy.

Formulate a Mathematical Model

It is convenient to develop the model of this process in stages as follows:

Stage 1: Consider the initial data communication of m values from the master transputer to the first slave in the chain. From our model assumptions, this transmission will take place in time

$$t = t_s + m\, t_c$$

Once this has occurred, two data transfers can take place in parallel, namely, m values passed from the master to the first slave again and m values from the first slave to the second slave. This passing-on of packets of data continues until the final transputer has received its m values. The total time for the distribution of the data is thus

$$t = (p-1)\,[t_s + m\, t_c]$$

Stage 2: The m values are summed by each transputer in time $m\, t_f$.

Stage 3: The slaves each pass back their result to the master in a manner similar to stage 1, except that only one value is being passed at a time. This is achieved in time

$$(p-1)\,[t_s + t_c]$$

Stage 4: The p partial sums now stored on the master are added and divided by n in time

$(p+1)\, t_f$ (i.e. p additions and 1 division)

The total time for parallel execution is thus

$$\begin{aligned} t_{par} &= (p-1)\left[t_s + mt_c\right] + mt_f + (p-1)\left[t_s + t_c\right] + (p-1)t_f \\ &= (m+p+1)\, t_f + (p-1)(m+1)t_c + 2(p-1)t_s \end{aligned}$$

To compare sequential performance with parallel processing, it is usual to calculate the ratio of the execution time on one processor divided by the execution time on p processors. This ratio is termed the speed-up and is given here by the expression*

$$S = \frac{(n+1)t_c}{(m+p+1)\, t_f + (p-1)(m+1)t_c + 2(p-1)t_s}$$

(Note: the numerator here is simply the time required for the sequential calculation of n additions and 1 division on the master transputer.)

Remembering that $n = mp$ and dividing numerator and denominator by t_f gives

$$S = \frac{(mp+1)}{(m+p+1) + (p-1)(m+1)c_1 + 2(p-1)c_2} \qquad (3.5)$$

where $c_1 = \dfrac{t_c}{t_f} \simeq 0.83$ and $c_2 = \dfrac{t_s}{t_f} \simeq 4.62$

Clearly, we would hope that S is at least greater than 1 (why?). DERIVE can be used to see how S changes as both m and p vary independently.

Interpretation

To interpret equation (3.5) using DERIVE, the reader is encouraged to proceed as follows:

(i) Declare variables m and p to have positive domain.
(ii) Author $C1 := 0.83$ and $C2 := 4.62$.
(Note: this has the effect of declaring the variables $C1$, $C2$ as well – see the *DERIVE User Manual*, Chapter 4).
(iii) Author the expression for S given by equation (3.5).

(iv) Investigate what happens to S as the size of the data set increases (i.e. as $m \to \infty$) by selecting Calculus Limit and typing INF for the limit point value. Simplify the resulting expression.

(v) Plot the resulting expression for S. (Choose a Scale of $x = 2$, $y = 0.25$ where x is synonymous with p and y with S. Centre the plot as appropriate – remember too that p is an integer ≥ 1.)

You should observe that the limiting value is greater than 1 for $p > 1$ and soon reaches a limit of around 1.2 as p increases.

(vi) Repeat steps (iv) and (v) but for the case $p \to \infty$ and m allowed to vary.

In each of the above cases a value for speed-up S greater than 1 is predicted, which is encouraging. However, the best we seem able to achieve is around $S = 1.2$, i.e. a 20% improvement. This figure will be even less for given finite values of m and p.

(vii) Investigate how S varies as m increases for a given value of p, say $p = 2$. Use Manage Substitute for p in your expression for S and Plot the graph. (suggested Scale x : 50, y = 0.25). Repeat using $p = 1, 5, 10, 15$, say, and superimpose all plots on the same axes.

You should observe that the plots for $p = 10$, 15 and beyond are almost identical, which suggests that in this case there is little to be gained by adding more and more processors to the chain to try to speed-up the calculation.

Use the cross-hairs to observe the value of S for various m and p values. Note, for example, that this model of performance predicts no speed-up (i.e. $S < 1$) for values of m less than around 65, no matter how many transputers (> 1), are used.

(viii) Repeat the investigation of S as in (vii) but this time as p increases for given values of m, say $m = 25, 50, 60, 100, 1000, 2000$. Observe that there is little change in the plots for $m > 1000$ and that for $m \leq 60$, no speed-up is predicted in agreement with the above result.

(ix) Now that some feel for the behaviour of S as p and m are varied independently has been established, try using DERIVE to plot S as a function of the two variables p and m. A typical plot obtained by the authors is shown in Fig. 3.6. Does your plot match your expectations? Is it easy to interpret? (Try varying the EYE position to vary the perspective and remember we are only interested in S for integer values of m and p each greater than unity.)

How good is this model of performance? A comparison of actual and predicted speed-up is reported in [4] and correlation is acceptable. Of course, this does not guarantee a successful prediction of performance if either the network topology is changed or a different algorithm and/or a different computation strategy is employed. The value of developing a model such as this is that it enables comparisons to be made between different algorithms and implementation strategies and is especially useful given the rate of change of transputer technology and its effect on the values of t_s, t_c and t_f.

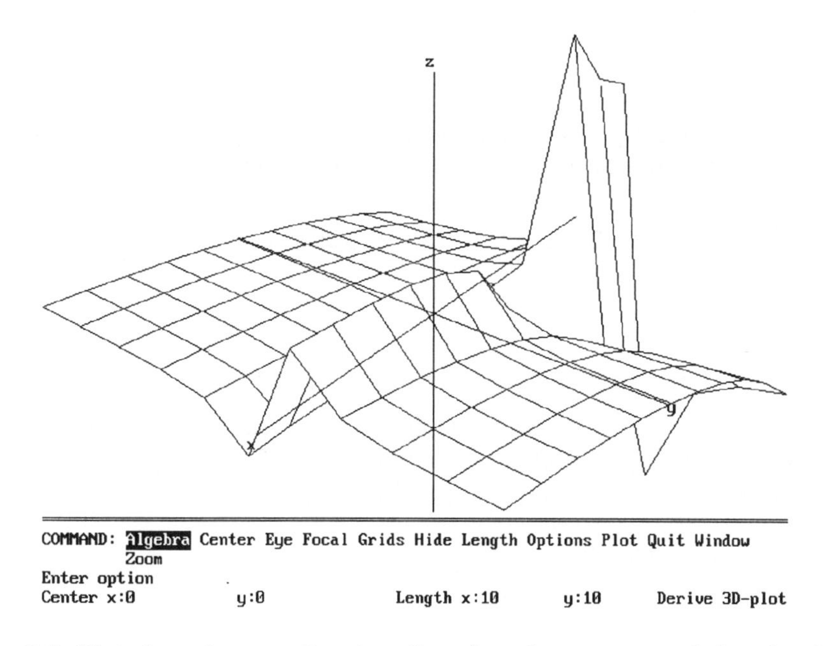

Figure 3.6 Plot of speed-up as a function of number of transputers and size of each data packet

Further Work

1. Using the same models for communication and arithmetic, establish a prediction for the speed-up S when calculating the mean $\overline{x}$ on a tree of transputers, as illustrated in Fig. 3.5(b). Here $p = 4$ and $m = n/4$ data items are sent to each branch in parallel with $n/4$ values kept on the root. The remaining stages are as in the previous study.

   Does your model predict a bigger speed-up? Intuitively, would you have expected this outcome or not?

   $$\left[\text{Solution} \quad S = \frac{(4\,m+1)}{(m+5) \; + \; (m+1)\,c_1 \; + \; 2\,c_2} \right]$$

2. Experiments conducted with a different type of transputer gave rise to measured performance constants as

$$t_s \simeq 12.2 \times 10^{-6}\ \text{s}$$

$$t_f \simeq 2.35 \times 10^{-6}\ \text{s}$$

$$t_c \simeq 4.4 \times 10^{-6}\ \text{s}$$

For a sample of $n = 60\,000$ data values, the following measured results were obtained for the calculation of $\bar{x}$ using a linear chain:

No. of processors, p	Computation time, t_{par} (s)	Measured speed-up, S
2	0.2039	0.691 51
3	0.2261	0.623 51
4	0.2372	0.594 57
5	0.2438	0.578 26

Use DERIVE to compute the model predictions for those measured values using equation (3.5) with new values for $C1$ and $C2$ and remembering that

$$m = \frac{n}{p} = \frac{60\,000}{p}$$

3. The standard formula

$$\bar{x} = \frac{1}{n}\sum_{i=1}^{n} x_i = M_1 \qquad \text{(say)}$$

for calculating the mean of a set of n values has been found to introduce serious accumulated round-off error under some circumstances (see reference [5]). An alternative algorithm proposed to overcome this error is the two-pass algorithm

$$M_2 = M_1 + \frac{1}{n}\sum_{i=1}^{n} (x_i - M_1)$$

To implement this on a linear chain, M_1 needs to be calculated as before, together with the following additional stages:

- Transmit the value of M_1 from the master to each slave transputer.
- Compute $\Sigma(x_i - M_1)$ on each transputer (assume that each transputer has retained a copy of its own m data values from the computation of M_1).

- Return the value of each partial sum to the master.
- Compute M_2 on the master.

Intuitively, do you feel that this is likely to lead to a greater speed-up? Compute the model prediction for speed-up and analyze the result.

4. A discerning reader might well have observed that the splitting up of the n data items into equal-size packets is not necessarily optimizing performance in that some processors are idle at some time in the process. For example, in the linear chain topology and with the strategy outlined, the master has to wait for the last slave in the chain to compute and return its sum before it starts to compute. There is nothing to stop the master starting to compute as soon as a partial sum is returned by a slave, i.e. stages 3 and 4 in the earlier work could be overlapped.

   Similarly, the packet size distributed along the chain could be so arranged that the master transputer retains a larger set of data values and the final slave only a few values so that the master finishes computing its own partial sum just as the slaves are ready to return theirs.

   Develop a model for the speed-up when such strategies are attempted to optimize performance. (Hint: start first with just two processors, a master and a slave, and suppose that the master keeps a fraction α of the data values and transmits a fraction $(1 - \alpha)$. Try to find a value for α that appears to minimize the parallel execution time for the calculation of $\bar{x}$. Once you have achieved this, try to generalize your model for 3, 4, 5 ... transputers.)

5. A computation problem common to mathematics is that of matrix–vector multiplication i.e. $A \cdot \mathbf{x}$ where A is an $m \times n$ matrix and $\mathbf{x}$ is an $n \times 1$ vector. Suppose that p processors of the type described in the case study are available. Use your modelling skills to advise

   - whether this problem is likely to benefit from the use of parallel processing,
   - if so, what topology should be used and what implementation strategy?
   - what assumptions about m, n, p, etc., are necessary for the proposed model?

(Hint: an obvious parallelization is to perform the inner product of each row of A with the vector $\mathbf{x}$ in parallel).

3.4 FOREST MANAGEMENT

The Problem

The authors are fortunate to live close to a heavily forested nature reserve, and while walking through the reserve one can observe a variety of trees, mostly pines, of varying heights ranging from newly planted saplings to fully mature trees. In more recent years, there has been more emphasis on forest management, with selected trees being felled and replaced by saplings. Forestry is now a commercial enterprise which offers financial returns to investors once the trees have been felled and the timber sold. Effective forest management is therefore important and should seek to:

(a) cut down enough trees at harvest time to maintain or extend profit margins,

(b) replace felled trees with saplings in such a way that the forest is not depleted but maintains its collection of trees at various stages of maturity so that regular profits can be made without damaging the ecology of the system.

A system model should attempt to select a harvesting strategy which will optimize the expected profit.

Setting up a Model

Features likely to be relevant to any model are as follows:

- the distribution of heights of trees in the forest at harvesting,
- the commercial value of a tree of a given height,
- the growth-rate of a tree (which may depend upon the height),
- the time interval between harvests (if any – harvesting may be continuous),
- the number of trees that die naturally (rather than being felled) in between successive harvests,
- the amount of growing space required by each tree and the total growing land available for the forest.

We shall consider a simple management model, as detailed in [6], which is based on the following assumptions:

- The area of land given over to forestry is fixed and any new growth has sufficient space to grow to full maturity if need be.
- All trees grow until harvesting, i.e. dying trees are ignored.

- Trees are classified by their height and hence at harvest time any given tree will fall into one, and only one, of the classes $\{h_1, h_2, \ldots, h_n\}$, where, for example,
  h_1 could be the class of all trees under 1 m tall,
  h_2 could be the class of all trees greater than or equal to 1 m tall but less than 2 m tall,
  $h_3 \ldots,$
  h_n could be the class of all trees greater than or equal to $(n-1)$ m tall.
- The period between harvesting is such that in this time, a tree belonging to class h_i at one harvest can move at most one class higher to h_{i+1} at the next harvest. (It could of course stay in its original class depending on the growth-rate.)
- The commercial value of a tree is related to its height and hence to the class h_i to which it belongs when felled.
- Any tree that is felled is replaced by a sapling, i.e. a tree belonging to the class h_1.
- The distribution of tree heights immediately after a harvest is the same for each harvest, to ensure that the growth–harvest cycle can continue.

Formulate a Mathematical Model

Firstly, let us define terms and units:

Let x_i $(i = 1, 2, \ldots, n)$ be the number of trees within class h_i after a harvest.

Let p_i $(i = 1, 2, \ldots, n)$ be the commercial value of a tree belonging to class h_i (£).

Let f_i $(i = 1, 2, \ldots, n-1)$ be the fraction of trees of class h_i that move into class h_{i+1} during the growth period between harvests. Hence $1-f_i$ will be the fraction that remain in class h_i.

Let r_i be the number of trees of class h_i felled at harvesting.

It is mathematically convenient to represent each of these subscripted variables as elements of column vectors, i.e.

$$\mathbf{x} = \begin{bmatrix} x_1 \\ x_2 \\ \vdots \\ \\ x_n \end{bmatrix}, \quad \mathbf{p} = \begin{bmatrix} p_1 \\ \vdots \\ \\ \\ p_n \end{bmatrix}, \quad \mathbf{f} = \begin{bmatrix} f_1 \\ \vdots \\ \\ \\ f_{n-1} \end{bmatrix}, \quad \mathbf{h} = \begin{bmatrix} h_1 \\ \vdots \\ \\ \\ h_n \end{bmatrix}, \quad \mathbf{r} = \begin{bmatrix} r_1 \\ \vdots \\ \\ \\ r_n \end{bmatrix}$$

Consider now a class h_i containing x_i trees. During the growth period, $(1-f_i)x_i$ of these will remain in this class and $f_{i-1}\, x_{i-1}$ will move into this class from the one below. This will be the case for $i = 2, 3, \ldots, n-1$ but for $i = 1$ we only have $f_1\, x_1$ trees growing out of this class while $(1-f_1)x_1$ remain, and for $i = n$, we have the number originally in the class, x_n, plus those growing into it $f_{n-1}\, x_{n-1}$ so that at the end of the growth period the distribution of the trees in the n classes is given by the vector

$$\begin{bmatrix} (1-f_1)x_1 \\ f_1\, x_1 + (1-f_2)x_2 \\ \vdots \\ \\ \\ f_{n-2}\, x_{n-2} + (1-f_{n-1})x_{n-1} \\ f_{n-1}\, x_{n-1} + x_n \end{bmatrix}$$

Now at harvest time, $r_1 + r_2 + \ldots + r_n$ trees are felled and replaced by trees of class h_1. Hence immediately after harvesting, the distribution of trees in the n classes is given by

$$\begin{bmatrix} (1-f_1)x_1 + r_2 + \ldots + r_n \\ f_1\, x_1 + (1-f_2)x_2 - r_2 \\ \vdots \\ \\ \\ f_{n-2}\, x_{n-2} + (1-f_{n-1})x_{n-1} - r_{n-1} \\ f_{n-1}\, x_{n-1} + x_n - r_n \end{bmatrix} \tag{3.6}$$

Since it is assumed that the distribution of tree heights immediately after a harvest is the same for each harvest, then

$$\begin{bmatrix} (1-f_1)x_1 + r_2 + \ldots + r_n \\ f_1\, x_1 + (1-f_2)x_2 - r_2 \\ \vdots \\ \\ f_{n-1}\, x_{n-1} + x_n - r_n \end{bmatrix} = \begin{bmatrix} x_1 \\ x_2 \\ \vdots \\ \\ x_n \end{bmatrix}$$

i.e.

$$
\begin{aligned}
r_2 + r_3 + \ldots + r_n &= f_1 x_1 \\
r_2 &= f_1 x_1 - f_2 x_2 \\
r_3 &= f_2 x_2 - f_3 x_3 \\
\vdots &= \vdots \\
&= \\
r_{n-1} &= f_{n-2} x_{n-2} - f_{n-1} x_{n-1} \\
r_n &= f_{n-1} x_{n-1}
\end{aligned}
\tag{3.7}
$$

Given a distribution x_i and known growth rates f_i, these n equations can (hopefully) be solved for the r_i. Once these are known, the revenue obtained can be computed as

$$Revenue = \sum_{i=1}^{n} p_i r_i$$

A closer examination of the system of equations (3.7) reveals that there are only really $n-1$ equations since the first one is the sum of all the rest. Also, r_1 does not appear in these equations. The problem statement would suggest that we take $r_1 = 0$ (why?). Thus we have $n-1$ equations for the $n-1$ unknowns. We still require to find the distribution x_i which will maximize the revenue.

Note also that since $r_i \geq 0$ ($i = 1, \ldots, n$), then the system of equations (3.7) leads to the following conditions on the initial distribution x_i namely,

$$f_1 x_1 \geq f_2 x_2 \geq f_3 x_3 \geq \ldots \geq f_{n-1} x_{n-1} \geq 0 \tag{3.8}$$

Also, since the size of the forest is fixed, we have the extra condition

$$x_1 + x_2 + \ldots + x_n = \text{constant } c \text{, say} \tag{3.9}$$

After eliminating the r_i, the problem formulation has been transformed to:

maximize $p_2 (f_1 x_1 - f_2 x_2) + \ldots + p_{n-1}(f_{n-1} x_{n-2} - f_{n-1} x_{n-1}) + p_n f_{n-1} x_{n-1}$

subject to the constraints (3.8) and (3.9)

This is known as a LINEAR PROGRAMMING problem and many text books cover the mathematical details of the subject (see, for example, [6]). See also the linear programming example in Chapter 4, page 78.

Solution of the Mathematical Model

For the particular case of a forest of Scots pine, the following values of f_i were measured for pines sorted into six classes (see [6]):

$$f_1 = 0.28\,,\ \ f_2 = 0.31\,,\ \ f_3 = 0.25\,,\ \ f_4 = 0.23,\ \ f_5 = 0.37$$

The financial value (in £) per tree for each of the six classes is assumed to be $p_1 = 0$, $p_2 = 10$, $p_3 = 20$, $p_4 = 30$, $p_5 = 50$, $p_6 = 80$ (saplings are of no value other than for planting; the tallest trees are sought after for public displays at festivities such as Christmas).

Suppose that, initially, the forest has 6000 trees divided equally among the six classes and that these are allowed to grow. Suppose also that the felling policy adopted is 'cut down all trees belonging to h_6 (i.e. the class of tallest trees) and relax the condition that the distribution of tree heights must stay the same'. What happens to the forest?

Here $\quad x_1 = x_2 = x_3 = x_4 = x_6 = 1000 \qquad$ (initially)

$$r_2 = r_3 = r_4 = r_5 = 0$$

$$r_6 = x_6 + f_5\, x_5$$

Mathematically, all that is needed is to start with the vector

$$\mathbf{x} = [\ 1000,\ 1000,\ 1000,\ 1000,\ 1000,\ 1000\ ]`$$

(where ` denotes the transpose of the vector) and then compute the new **x** vector given by expression (3.6) using the appropriate f values. This process is then repeated to generate successive **x** vectors representing the distribution of trees in each class after successive harvests.

This tedious, iterative calculation can be more easily performed by using DERIVE if expression (3.6) is re-written in the equivalent form

$$G\,\mathbf{x} = \begin{bmatrix} 1-f_1 & 0 & 0 & 0 & f_5 & 1 \\ f_1 & 1-f_2 & 0 & 0 & 0 & 0 \\ 0 & f_2 & 1-f_3 & 0 & 0 & 0 \\ 0 & 0 & f_3 & 1-f_4 & 0 & 0 \\ 0 & 0 & 0 & f_4 & 1-f_5 & 0 \\ 0 & 0 & 0 & 0 & 0 & 0 \end{bmatrix} \begin{bmatrix} x_1 \\ x_2 \\ x_3 \\ x_4 \\ x_5 \\ x_6 \end{bmatrix}$$

The repeated calculation of $G\mathbf{x}$ for successive vectors $\mathbf{x}$ can now readily be performed using the ITERATES function as follows:

Author **x**0 := [1000, 1000, 1000, 1000, 1000, 1000]`

$$G := [\,[0.72, 0, 0, 0, 0.37, 1], [0.28, 0.69, 0, 0, 0, 0], [0, 0.31, 0.75, 0, 0, 0], [0, 0, 0.25, 0.77, 0, 0], [0, 0, 0, 0.23, 0.63, 0], [0, 0, 0, 0, 0, 0]\,]$$

where the values of f_i specified earlier have been used.

Then Author

iterates (G.**x**, **x**, **x**0, 10)

where the 'dot' signifies the dot product of the matrix G and the vector $\mathbf{x}$, (see *DERIVE User Manual*, Chapter 8) and the value 10 indicates the number of iterations completed (i.e. harvest periods). ApproXimate this expression.

Interpretation

The resulting sequence of vectors shows that the distribution of trees in each class changes after each harvest, certainly for the first few harvests. However, using Manage, Substitute to vary the number of iterations to, say, 50, 100, 200 and re-computing the sequence of vectors yields the result that the distribution eventually settles down to a 'steady-state', i.e. the distribution doesn't change after further harvests.

Thus we are predicting that this felling policy is sustainable in the long term. Note, though, that the revenue from this policy falls initially before rising to a steady income – and we have not considered the cost of replacing the felled trees with saplings (or their availability)!

Is this convergence to a steady value for $\mathbf{x}$ dependent upon the choice of initial distribution $\mathbf{x}_0$? Try the calculations again with any $\mathbf{x}_0$ of your choice, provided that the sum of the elements is 6000. You should observe that a steady distribution is eventually reached in each case with the same steady-state vector $\mathbf{x}$ being achieved.

Further Work

Readers with some knowledge of matrices may well have observed that the matrix G is an example of a MARKOV MATRIX since all the elements have values in the range [0, 1] and the elements of each column sum to 1. (G is not a STRICT MARKOV MATRIX since some elements are zero.) Such matrices have a property that the sequence

$$\mathbf{x}_{i+1} = G\,\mathbf{x}_i \qquad i = 0, 1, 2, \ldots$$

will converge to a limiting value that is an eigenvector of G corresponding to an eigenvalue $\lambda = 1$ of G. This result can be confirmed in this case by using the function APPROX_EIGENVECTOR (G, 0.999 99) in the utility file VECTOR.MTH (see *DERIVE User Manual*, Chapter 9) and observing that the resulting eigenvector is proportional to the steady-state $\mathbf{x}$ obtained earlier.

DERIVE has some difficulty with this 6×6 matrix G when in exact mode. The DERIVE User Manual suggests that for matrices of this size, approximate mode should be used. The eigenvalues of G can be found by using CHARPOLY(G) to determine the characteristic polynomial of G and then plotting this expression to observe that $\lambda = 1$ is an eigenvalue.

APPROX_EIGENVECTOR has also been used to determine the eigenvector corresponding to $\lambda = 1$, but note that an estimate 0.999 99 for the exact eigenvalue $\lambda = 1$ has been used in the arguments list, as recommended in the *DERIVE User Manual*. Interested readers may wish to vary this estimate to (say) 0.9, 0.99, etc., or even try a value 1.0 to observe the influence of this argument on the output from the utility function.

The linear programming problem presented in the case study has been shown in [6] to have the following solution:

> The optimal revenue is achieved by harvesting all of the trees from one particular height class and none of the trees from any other height class.

In the notation adopted here, this optimal revenue is given by the largest value of

$$\frac{p_k \cdot C}{\dfrac{1}{f_1} + \dfrac{1}{f_2} + \dots + \dfrac{1}{f_{k-1}}}, \qquad k = 2, 3, \dots$$

where the value of k which maximizes this expression is the number of the class which is completely harvested.

For the problem in the previous exercise, $C = 6000$ (the total number of trees), use DERIVE to determine which class should be completely harvested to yield the optimum revenue in this case. (You should find that felling the class of tallest trees yields the optimum revenue in this case.)

Suppose that market forces adversely affect the demand for the tallest trees and that consequently the price p_6 falls from £80 per tree to £60. What felling policy would you now advise?

A different forest is similarly classified and is found to have the distribution

$$h_1 = 500, \quad h_2 = 300, \quad h_3 = 200, \quad h_4 = h_5 = 100, \quad h_6 = 0$$

The management policy is merely one of felling all trees that lie in class h_6 (at the next harvest cycle and beyond), without any re-planting or concern for the forest's future. Assuming the same growth-rates as before, determine whether or not this forest will eventually be totally depleted and if so how many harvesting periods this will take.

3.5 MODELLING THE BRAIN AND ITS BEHAVIOUR

The Problem

Attempts to model the nervous activity in the brain, and hence to model human thought processes, have been undertaken by biologists and psychologists since before the turn of this century and have given rise to the term 'neural networks'. Since the 1970s research activity into the development of neural networks has increased hand in hand with advances in computer technology, so that nowadays neural network development attracts both biologists and computer scientists. Here we examine how simple network models have been developed. The interested reader is referred to [7] and [8] for more details of the basic ideas and novel applications.

Setting up a Model

Clearly, some initial knowledge of the biological features of the brain is needed. The most common models are based on the NEURON as the basic feature. Fig. 3.7 shows a neuron composed of a body, one AXON which may be thought of as a long thin tube which splits into branches and DENDRITES which form a very fine 'brush' surrounding the body of the neuron. The branches at the end of the axon have ENDBULBS which almost touch the dendrites of other neurons, the small gap between them being called a SYNAPSE.

Impulses move down the axons of a neuron and interact with the dendrites of another neuron via the synapses. The strength of the impulse signal passed on depends upon the efficiency of the synaptic transmission. A given neuron will send an impulse down its axon if it has received sufficient signals from other neurons in a short period of time and if the strength of those signals is greater than a critical amount known as the bias. The signal impinging on a dendrite is said to be either excitatory or inhibitory.

Armed with this biological description, what are the modelling possibilities? We shall develop a model based on the following three assumptions:

- The neuron can be considered as a unit that receives signals from other neurons.
- If x_1, x_2, ..., x_j ,... represent the other neurons connected to neuron x_i ($j \neq i$), then the efficiency of the synaptic transmission between neuron x_i and x_j can be approximated by a weight w_{ij}, where w_{ij} can be positive (excitatory) or negative (inhibitory).

LIVERPOOL JOHN MOORES UNIVERSITY
LEARNING SERVICES

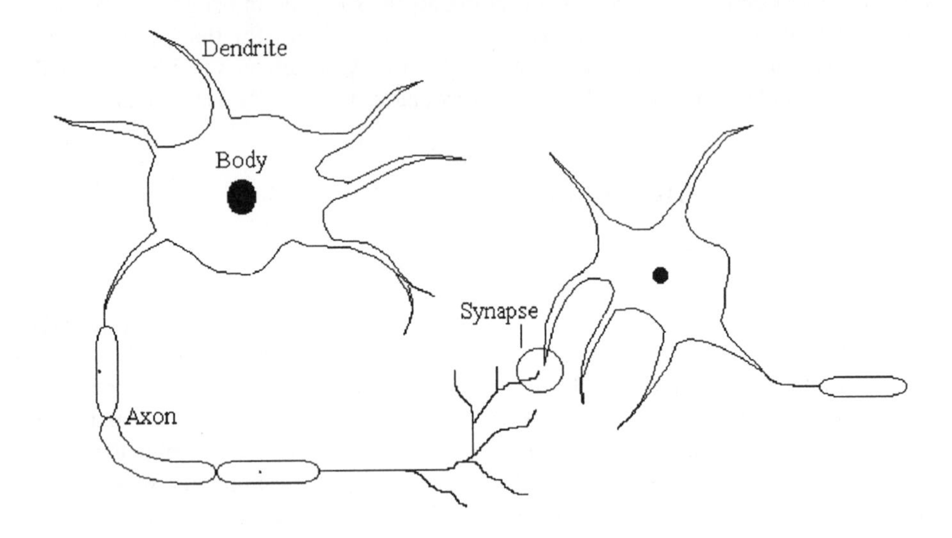

Figure 3.7 A neuron

- The bias term for neuron x_i can be approximated by a constant term β_i. (The bias term may vary from one neuron to another but is a constant β_i for each neuron x_i.)

Formulate a Mathematical Model

If we assume that the inputs to a neuron are summed linearly (following the usual advice to 'keep it simple'), then we can model the activity level of neuron x_i by

$$\eta_i(t) = \sum_j w_{ij}\, x_j(t) + \beta_i \qquad (j \neq i) \tag{3.10}$$

In a neuron x_i, it seems reasonable to suppose that the output activity (or firing rate) y_i will be some function of the input activity level, i.e.

$$y_i(t) = f(\eta_i(t)) \tag{3.11}$$

The above is summarised diagrammatically in Fig. 3.8. Different choices for the function f result in different models for the neuron.

Of course the brain has countless numbers of such neurons and associated connections, whereas any model is likely to be limited by size. Even so, modelling the behaviour of a network of such neurons remains a challenge for the future.

Many models have been proposed for the function $f(\eta)$. For example,

(a) $$f(\eta) = \begin{cases} \text{ON} & \text{if } \eta > \text{ some threshold value} \\ \text{OFF} & \text{if } \eta < \text{ this threshold value} \end{cases}$$

such a neuron is termed a PERCEPTRON.

(b) $$f(\eta) = \eta$$

such a neuron is termed an ADALINE.

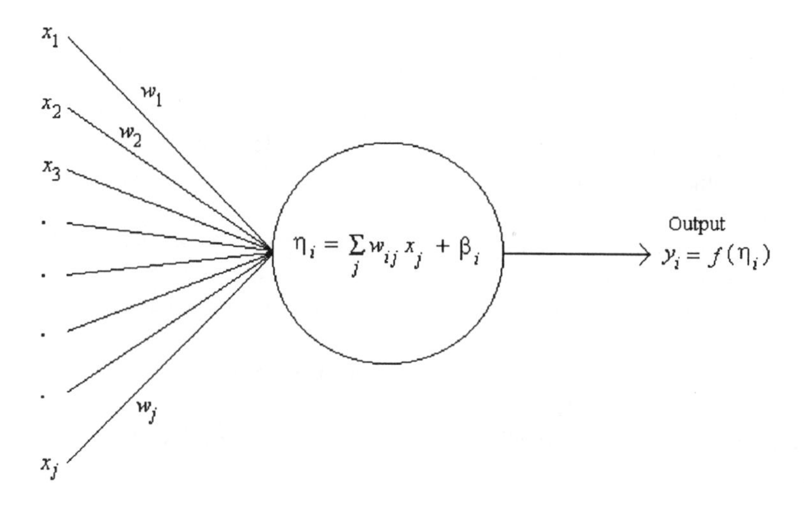

Figure 3.8 A mathematical model of a neuron

The most popular choices are SIGMOIDAL functions which have the properties that they are monotonically increasing, continuous, continuously differentiable and have asymptotic values (usually 1 and 0) as η tends to $+\infty$ and $-\infty$ (respectively).

Two functions which possess these properties are:

$$f(\eta) = \frac{1}{1 + e^{-\eta/T}} \quad \text{and} \quad f(\eta) = \frac{1}{1 + e^{-(\eta+T)}}$$

where T is some threshold value.

However, postulating forms for $f(\eta)$ does not complete the formulation. The weights w_{ij} for a given network of connections are not known and are generally estimated by 'training the network'. Known inputs are put onto the network with assumed values for the weights and the computed outputs compared with the known outputs. The weights are adjusted until computed and known outputs are in reasonable agreement.

Solution of the Mathematical Model

Use the DERIVE plotting facilities to convince yourself that the sigmoidal functions quoted above do indeed have the required characteristics. Try to write down other possible functions that you think may have this property. (Assume $T = 1$ for simplicity.) Since the simplest model computes the sum $\sum_j w_{ij} x_j$ (assuming β_i is zero), and since for a fixed i this is simply the scalar product of an input vector **x** and an associated weight vector, **w**, the orthogonality property of vectors can be used to good effect to produce a neural network capable of simple pattern recognition.

An image can be represented on a computer by an ordered sequence of binary digits and hence an image can be thought of as a pattern of 0s or 1s. Suppose **x** holds the array of binary digits to be tested and **w** holds a known pattern. If **x** and **w** are normalized (i.e. each element is divided by the magnitude of the vector), then **x.w** will lie between 0 and 1 inclusive. The bigger the value of **x.w** then the closer **x** and **w** match each other.

Suppose that your model consists of n input neurons each containing a 1 or a 0 so that the n neurons taken in order generate a bit pattern and that a number of ADALINE neurons are connected to each input neuron (but not to each other). (Assume the bias β is zero in each case.) For each ADALINE, fix the n weights as you wish so that they, too, form a bit pattern. For a normalized input pattern **x** of your choice and for each of the normalised weight vectors (or pattern testers) **w** associated with each ADALINE, compute **x.w** and confirm that this simple network does recognize an input pattern.

References

[1] Fox M., 'Formulae funding of schools: some mathematical results'. *Mathematical Gazette*, 77, (480), 337–352 (1993).

[2] INMOS Ltd, *Transputer Reference Manual*, Prentice Hall, (1988).

[3] *Transputer Technology and its Applications – A distance-learning video*, Liverpool John Moores University (1992).

[4] Clegg D., Al-Jumeily D., Pountney D.C. and Harris P., 'On the modelling of simple statistical calculations using distributed memory computers'. *Liverpool John Moores University, School of Computing and Mathematical Sciences, Internal Report* (1994).

[5] Neely P.M., 'Comparison of several algorithms for computations of means, standard deviations and correlation coefficients'. *Communications of the ACM*, **9**, July (1966).

[6] Rorres C. and Anton H., *Applications of Linear Algebra*, J. Wiley (1994).

[7] Rumelhart D.E., Widrow B. and Lehr M.A., 'The basic ideas in neural networks'. *Communications of the ACM*, **37** (3), 87–92 (1994).

[8] ibid. 'Neural networks: Applications in industry, business and science'. *Communications of the ACM*, **37** (3), 93–105 (1994).

4

Optimization Models

4.1 INTRODUCTION

This chapter is about optimization. It is probably true to say that when they think mathematically about optimization, most people think in terms of calculus and maximum and minimum values.

While there are certainly many optimization problems which fit into this category (and we shall look at some in this chapter), there are also many for which the calculus approach is inappropriate. Among these are, for example, some of the problems relating to the optimal management of resources and the optimization of profits/minimization of costs of complex manufacturing processes. Such problems require the techniques of linear programming for their solution. While DERIVE does not possess specific linear programming routines, it can nevertheless be used with some problems to illustrate the concepts behind linear programming and to solve some simple linear programming problems.

As this is a topic which is probably less familiar than calculus to the majority of our readers, we begin this chapter with an example whose solution is posed as a linear programming problem.

4.2 SCHEDULING OF DOCKSIDE LOADING FACILITIES

The Problem

The demand for coal, iron ore, etc., has led to the construction of very large bulk carriers to transport the material by sea. These vessels are often so large that they may be restricted to entering harbours near low water and leaving near high water (since their draught is so great once they are loaded).

A certain port authority has two berths for such vessels and three loading machines (see Fig. 4.1). Develop a model of the loading system which can be used by the harbourmaster to optimize the shiploading schedules. The solution presented here is based on an article by W. Galvin, [1].

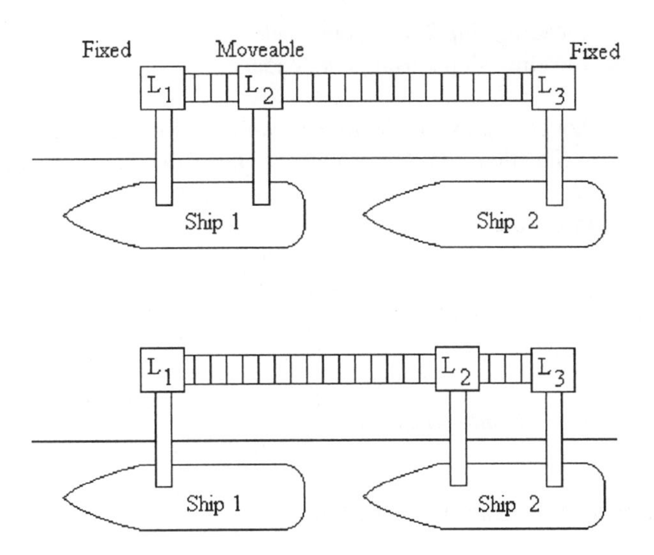

Figure 4.1 Loader 1 can only service berth 1, with ship 1,
Loader 2 can service either berth,
Loader 3 can only service berth 2, with ship 2.

Setting up a Model

We shall begin by assuming that empty ships enter harbours at the time of low water ($t = 0$) and that high water occurs six hours later. Rather than allow the full six hours for loading it would be prudent to allow some time for delays, repairs to the loaders, etc., say two hours. It then follows that all loading must be completed four hours, ten hours, sixteen hours later in order that laden ships can leave at high water.

The loading of each ship will depend on her capacity, the rate of loading and whether she is being loaded with one or two loaders.

The harbourmaster needs a model which can be used to predict the time at which he or she might move loader L_2 from one ship to the other in order to ensure that both ships are fully laden and ready to leave on the next high tide.

Formulate a Mathematical Model

Suppose that at time $t = t_0$ two ships are being loaded, with two loaders working on ship S_1 and one on ship S_2. Let T_1 and T_2 (tonnes) denote the tonnage already loaded on each ship at time $t = t_0$ and let K_1 and K_2 denote the capacity of each ship.

Let
R_{11} = rate of loading ship 1 using one loader (tonnes per hour),
R_{12} = rate of loading ship 1 using two loaders,

R_{21} = rate of loading ship 2 using one loader,
R_{22} = rate of loading ship 2 using two loaders.

Suppose that $t = t_C$ (hours) represents the time at which the moveable loader is changed over from one ship to the other, that the changeover takes one hour and that all loading is completed by time $t = t_F$.

The two loading rates are then

for ship 1, *Loading rate* $= \begin{cases} R_{12}\ , t_0 \le t \le t_C \\ R_{11}\ , t_C \le t \le t_F \end{cases}$

for ship 2, *Loading rate* $= \begin{cases} R_{21}\ , t_0 \le t \le t_C + 1 \\ R_{22}\ , t_C + 1 \le t \le t_F \end{cases}$

The total tonnages loaded at each berth are then

at berth 1 : $T_1 + R_{12}\,(t_C - t_0) + R_{11}\,(t_F - t_C)$

at berth 2 : $T_2 + R_{21}\,(t_C + 1 - t_0) + R_{22}\,(t_F - t_C - 1)$

Therefore, since both ships are required to be fully laden by time $t = t_F$ we must have

$$T_1 + R_{12}\,(t_C - t_0) + R_{11}\,(t_F - t_C) \ge K_1 \tag{4.1}$$

and

$$T_2 + R_{21}\,(t_C + 1 - t_0) + R_{22}\,(t_F - t_C - 1) \ge K_2 \tag{4.2}$$

Solution of the Mathematical Model

The striking feature of the mathematical problem represented by (4.1) and (4.2) is that it is posed in terms not of the usual equations but of inequalities and we want to solve these two inequalities for t_C (the time of the loader changeover). In order to see clearly how to progress from here (and to obtain a numerical solution for the harbourmaster), we need some data. The following ship capacities and loading rates may be considered representative:

$K_1 = 90\,000$ tonnes

$K_2 = 50\,000$ tonnes

(obviously in the scenario presented here it is sensible that $K_1 > K_2$; why else would you start with two loaders serving ship 1?)

$R_{11} = 2000$ tonnes/hour
$R_{12} = 3500$ tonnes/hour
$R_{21} = 2000$ tonnes/hour
$R_{22} = 3500$ tonnes/hour

Typical values for t_0 and t_F might be two hours and sixteen hours respectively. The value of t_0 has been selected at random but the value for t_F represents the possibility of departure on the second subsequent high tide (eighteen hours after entry) less a two-hour allowance for delays and repairs.

The inequalities (4.1) and (4.2) then become

$$T_1 \geq -1500\,t_C + 51\,000 \tag{4.3}$$

$$T_2 \geq 1500\,t_C - 500 \tag{4.4}$$

which are to be solved for t_C subject to the three constraints

$$T_1 \leq 90\,000\,, \quad T_2 \leq 50\,000 \tag{4.5}$$

and

$$2 \leq t_C \leq 16 \tag{4.6}$$

When the equalities contained in (4.3)–(4.6) are plotted using DERIVE, and regard is given as to whether the constraints are 'greater than' or 'less than' inequalities, two specific regions (called the feasible regions) of the capacity-time plane are defined (one for each ship), see Fig. 4.2. We have used the Centre facility in order to use the display area to maximum effect (only the positive quadrant is required as both the capacity and the time are greater than or equal to zero). The reader is encouraged to identify the feasible regions for each ship in Fig. 4.2.

DERIVE is unable to plot the vertical boundaries by simply Authoring the equation $t = 2$ (or 16). Instead, it is necessary to specify the equation of the line parametrically as

$$[2, a] \qquad \text{or} \qquad [16, a]$$

and then at the Plot command change the minimum and maximum value of a to 0 and 90 000 (the minimum and maximum capacities).

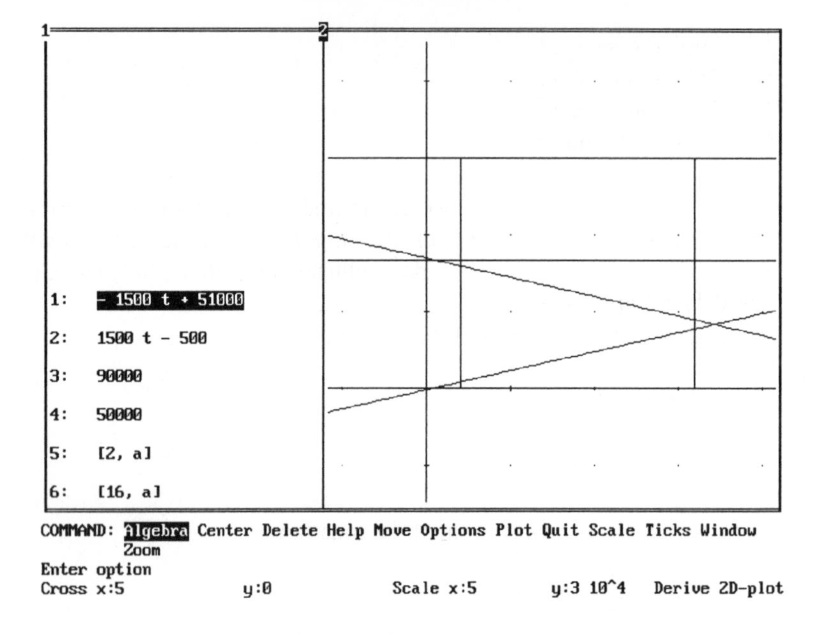

Figure 4.2 Graphs of the constraint equations for identification of the feasible regions

Interpretation and Validation

Suppose that at time $t = t_0$ (i.e.2 hours) ship 1 has 45 000 tons on board (i.e. $T_1 = 45\,000$) and ship 2 has 14 500 tons on board (i.e. $T_2 = 14\,500$), then Fig. 4.2 shows that

for ship 1, $t_C \geq 4$

for ship 2, $t_C \leq 10$

These two inequalities are consistent, so, provided that the moveable loader is changed over from ship 1 to ship 2 anytime between $t = 4$ hr and 10 hr after the initial low water then both ships can leave, fully loaded, on the second high tide. More complicated dockside loading facilities can be analyzed in a similar way.

Further Work

1. You could investigate this problem for other sizes of bulk carrier, different loading rates and different values of t_0.
2. Investigate the scheduling of three loading machines on to three berths, as shown in Fig. 4.3, with the same objective as that described earlier.

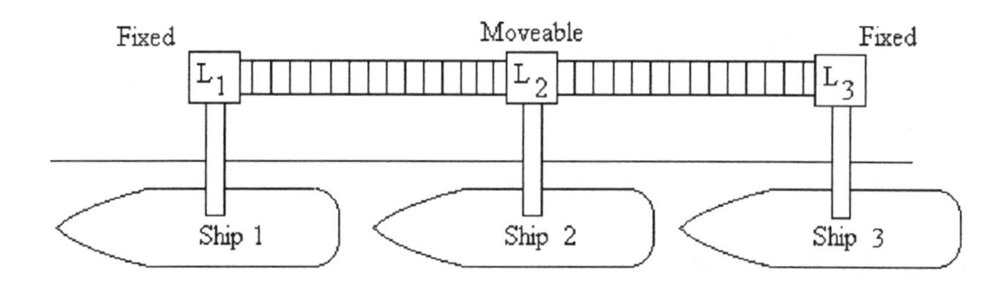

Figure 4.3 Alternative loading arrangement - three moveable loaders serving three ships

Optimization problems involving more than two independent variables cannot be simply handled using DERIVE as it is not possible to represent the inequalities graphically in more than three dimensions. Such problems can, however, be handled algebraically using a technique known as the Simplex method. The interested reader is referred to [2], which contains a clear exposition of the method.

4.3 LOTTERIES AND RAFFLES

Introduction

Lotteries are a common way of raising money for worthy causes. They are widely used by groups of all sizes ranging from national governments (who organize national lotteries) to local youth clubs/churches (who may organize a small lottery/raffle as a fund raising event).

Whatever the size of the venture, the basic aim is the same: to make as large a profit as possible for the organization running the event. This is achieved by making the prize(s) so attractive and the ticket price so reasonable that large numbers of people will be persuaded to take part. Too high a ticket price will deter some people from taking part,as will a prize which is not considered sufficiently attractive.

The Problem

Here we shall consider the case of a student society which organizes a raffle within a university in order to generate some funds for the society. The committee has decided to offer a single prize, which will have to be paid for out of the raffle ticket sales. The committee's previous experience has been that as the ticket cost increases there is a decrease in the number sold.

Setting up a Model

It is required to develop a model of such a raffle from which the optimal ticket cost and prize value, which maximize the profit of the venture, can be determined. The profit may be defined as

$$profit = income\ raised\ from\ ticket\ sales - overheads \tag{4.7}$$

We shall assume the following:

- The greater the ticket sales revenue the more valuable can be the prize offered.
- As the cost of the ticket increases, the number of tickets sold decreases.
- People only buy one ticket each.
- The tickets are printed by a wellwisher (and hence do not need to be paid for).
- Ticket distribution/sales is done freely by club members.

Formulate a Mathematical Model

Next we define the variables to be used and then use them to express our model (see equation (4.7)) in terms of them.

Let

n = number of tickets sold
C_T = cost of one ticket (pence)
C_P = cost of the prize (pence)
P = total profit made (pence)
S = ticket sales revenue (pence)

Equation (4.7) may then be expressed mathematically as

$$P = S - C_P$$

i.e.

$$P = n\,C_T - C_P \tag{4.8}$$

as the only cost is that of the prize, members having provided free printing and distribution.

In order to progress any further we now need submodels for the relationship between ticket cost (C_T) and number sold (n) and between ticket revenue and prize value (C_P). The simplest model for either of these relationships is a linear model, although in reality they are more likely to be non-linear. Suitable linear (and non-linear) forms are shown in Fig.4.4.

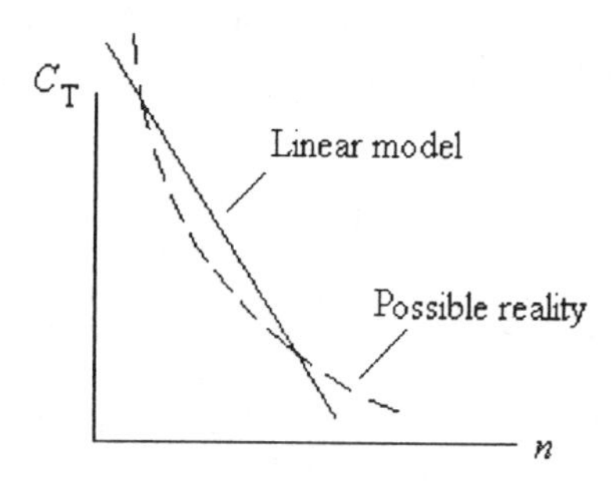

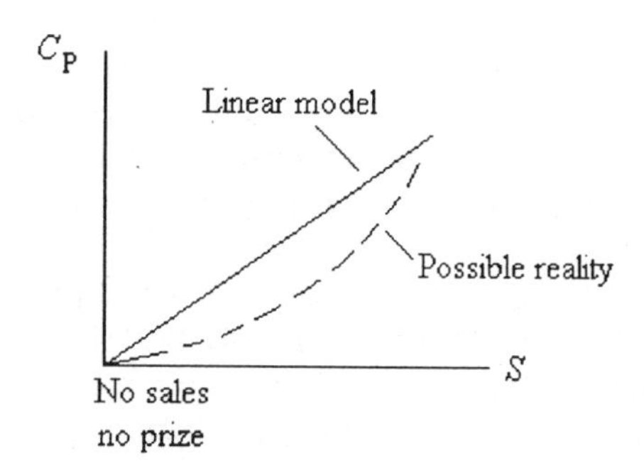

Figure 4.4 Models showing effect of ticket sales on ticket price and value of prize

Solution of the Mathematical Model

To quantify these two linear models some data are required. A sample poll among 200 students indicated that while 200 students would pay 10p for a ticket for a £5 prize, only 100 students would pay 50p for a ticket for a £25 prize. It is left as an exercise for the reader to show that the straight lines of Fig. 4.4 can then be expressed as

$$C_T = 90 - 0.4\,n$$

and

$$C_P = 2S/3 - 833$$

The profit expression is then finally expressed in terms of the number of tickets sold (n) as

$$P = 833 + 30\,n - 0.13\,n^2$$

Use the DERIVE Calculus facility to determine the first and second derivatives of P with respect to n and hence confirm that P has a maximum value of £25.64 when the ticket cost is 44p.

Interpretation/Suggested Further Work

On the basis of the data given, the organizers should sell the tickets at 44p each to realize a profit of £25.64.

The ticket price is somewhat awkward (all that change!) and suggests that the model needs some refinement. What would happen if the ticket price were reduced to 40p or, alternatively, raised to 50p for convenience?

The model could be revised by collecting additional data relating to the likely number of tickets sold at different prices for different values of prize. You could obtain such data by conducting a poll among your fellow students. The two graphs shown in Fig. 4.4 would then probably be non-linear. Curve fitting techniques could be employed to establish the nature of these two relationships and the expression for the profit (P) could then be revised. It could then be optimized as before.

One (at least!) of the original modelling assumptions was unreasonable, namely, the assumption that people only bought one ticket each. In practice, people often buy more than one ticket. For example, if the ticket price had been set at the more convenient 40p, then they might have been offered at 40p each or six for £2. What effect might multiple purchases have on the venture, from the organizer's point of view (obviously the purchaser's chances of winning are improved)?

4.4 OPTIMIZING SPORTS PERFORMANCES

Introduction

Sport is an obvious area in which optimization plays a key role whether it be maximizing the range of a throw in a field event or optimizing the strategy to be adopted in a team game.

The Problem

In this section we look at the role of optimization in the shot putting event. In the absence of air resistance, wind and spin effects (is it reasonable to ignore these in this event?) the range, R(m), of the throw is dependent on the release velocity, V(m s^{-1}), the release height, h(m), and the release angle with respect to the horizontal, α (degrees), and may be expressed as, see reference [3],

$$R = \frac{V^2 \cos\alpha}{g}\left\{\sin\alpha + \sqrt{\left(\sin^2\alpha + \frac{2gh}{V^2}\right)}\right\}$$

Through coaching, the thrower can change any of the three parameters V, α or h either by strength training or technical adjustments to the style of throwing. The obvious question to ask in the quest for enhanced performance is whether any one of these parameters is significantly more important than the others?

Formulate a Mathematical Model

For example, suppose we consider the angle of release. It is well known that the optimum angle of release for projectiles released from ground level is 45°. Is this also the case when the implement is released from a height h above the ground? We can investigate this by constructing a table of values of R for different values of α, for fixed values of V and h. (For élite senior shot putters appropriate values are in the ranges V: 10–14 m s^{-1}, α = 40°–45°, h = 1.8–2.5 m.) You can accomplish this with DERIVE by first Authoring the expression for R (remember that an angle A multiplied by DEG permits you to enter the angle A in degrees) and then using the Manage, Substitute commands to insert the values $g = 9.81$, $V = 13.7$, $h = 2.25$ so that R is then only dependent on the angle A (representing α). Since we are investigating whether 45° is significant for this problem use the VECTOR function to produce a list of values of R, for A ranging from 40° to 50° in steps of 1°. You will need to Author and then approXimate the function

VECTOR (#*n*, *A*, 40, 50)

where expression n contains the formula for the range R, after insertion of the V and h values. Your results should indicate that the range is a maximum when the shot is released at an angle within the range 40–45° to the horizontal. You could experiment further to satisfy yourself that 45° no longer has the significance it had for release at ground level.

Solution of the Mathematical Model

If 45° is not the optimum release angle, then what is? Your table should suggest 42° but it can also be determined precisely by first differentiating the expression for R with respect to α and then solving the equation

$$\frac{dR}{d\alpha} = 0$$

The interested reader is referred to [3], but the algebra is tedious. A more acceptable approach is to enlist the assistance of DERIVE as follows.

First, Declare α, g, h and V to be non-negative variables and then Author the expression for the range R:

$$R = \frac{V^2}{g}\cos\alpha\left\{\sin\alpha + \sqrt{\left(\sin^2\alpha + \frac{2gh}{V^2}\right)}\right\}$$

The above declaration merely ensures that the subsequent optimization does not contain terms such as sgn (h), sgn (V), etc., since each such term is known to be equal to unity as their arguments are each positive.

Use the Calculus and soLve facilities to show that the optimal value of the release angle α is given by the solution of the equations

$$\sqrt{\left(V^2\sin^2\alpha + 2gh\right)} + V\sin\alpha = 0$$

which actually has no solutions since all the variables involved are positive and $0° < \alpha < 90°$, and

$$V\cos^2\alpha - \sin\alpha\sqrt{\left(V^2\sin^2\alpha + 2gh\right)} = 0$$

By rearranging the above into the form

$$V\cos^2\alpha = \sin\alpha\sqrt{\left(V^2\sin^2\alpha + 2gh\right)}$$

and then squaring both sides (to remove the square root sign) the optimal value of α is given by solving

$$V^2\cos^4\alpha = \sin^2\alpha\left(V^2\sin^2\alpha + 2gh\right)$$

Direct solution of this equation for α is virtually beyond DERIVE's current capability. In fact, the use of DERIVE here is something of an overkill. Some thought should suggest that the substitution $y = \sin^2\alpha$ might be helpful. Its implementation leads directly to the result

$$\sin\alpha = \frac{\sqrt{2}V}{2\sqrt{\left(gh + V^2\right)}}$$

The value of the release angle α which maximizes the range of the throw is then given by

$$\alpha = \operatorname{asin}\left(\frac{\sqrt{2}\,V}{2\sqrt{\left(gh + V^2\right)}}\right)$$

Author the above expression for α and then use Manage and Substitute to insert the values used earlier for g, h and V to confirm the outcome of your earlier numerical experiments. Note that in the case where $h = 0$ (release from ground level) the 45° result is obtained. This serves as something of a check on the above result.

Further Work

Rather than concentrating solely on one of the parameters, as we have just done with the release angle, another worthwhile investigation is to establish which of the three parameters has the greatest effect on the range of the throw. This is left as a directed exercise for the reader, one suggested approach being based on the approximation

$$\delta y \simeq \frac{\mathrm{d}y}{\mathrm{d}x} \cdot \delta x$$

This can be adapted to deal with R, which is a function of the three variables α, h and V, as follows:

First, decide upon a representative value for each of α, h and V and determine the corresponding value of R, using Manage and Substitute. Again with Manage and Substitute, insert your chosen values for two of the variables (say α and h) into the original expression for R to produce a function of the single variable V.

Next, use Calculus, Differentiate to determine $\mathrm{d}R/\mathrm{d}V$, followed by Manage and Substitute to obtain its value for your chosen value of V.

To investigate the effect of changing the value of V suppose we make a 1% change so that

$$\delta V = \pm 0.01\, V$$

Evaluate this for your chosen value of V and hence obtain the change in the range by calculating

$$\delta R \simeq \frac{\mathrm{d}R}{\mathrm{d}V} \cdot \delta V$$

Repeat the procedure for changes in either α or h and examine the value of the small changes in the value of the range.

Interpretation and Validation

Your results should confirm the advice, given in the coaching manuals, that throwers should concentrate on achieving a high release velocity for the shot in order to maximize the distance thrown.

This result might have been anticipated given the presence of the terms V^2 in the expression for the range.

4.5 TIN CAN MANUFACTURE

The Problem

Companies involved in the production of tin cans for soft drinks, pet foods, etc., are concerned to minimize their production costs, commensurate with producing a tin can which their customers will want to buy.

Setting up a Model

One obvious feature to minimize would be the amount of metal used in the production of a can. We shall develop a model which addresses this problem and which is based on the assumption that the can is cylindrical and of circular cross-section (the presence of a rim and a concave base is ignored; is this reasonable?) and that its capacity is the standard 330 ml.

Formulate a Mathematical Model

Denoting the radius of the circular base by r (cm) and the height by h (cm) the surface area of metal required S (cm^2) is given by

$$S = 2\,\pi r h + 2\,\pi r^2 \tag{4.9}$$

This expression is to be minimized with respect to r and h and any solution must also satisfy the capacity requirement that

$$\pi r^2 h = 330 \tag{4.10}$$

This latter equation is known as a constraint equation. The presence of such equations is common in the mathematical formulation of practical optimization problems.

Solution of the Mathematical Model

The incorporation of constraints into the mathematical solution of the problem is achieved either by elimination (as we shall demonstrate in this example) or by adopting the method of Lagrange multipliers (as is used in other examples and is also presented here for comparison). It is possible to eliminate h between equations (4.9) and (4.10) and obtain an expression for S purely in terms of r as

$$S = \frac{660}{r} + 2\pi r^2 \tag{4.11}$$

It is left as an exercise for the reader to Author the function of r expressed in equation (4.11) and then use the Calculus facility to verify that S has a minimum value corresponding to

$$r = 3.745 \text{ cm} \qquad \text{and} \qquad h = 7.490 \text{ cm}$$

(It should be noted that in the majority of such optimization problems it is not generally possible to conveniently perform the algebraic elimination indicated above.)

Interpretation and Validation

How do these results compare with the actual dimensions of standard 330 ml drinks cans? Why is there a difference? After all, the mathematical solution represents minimum usage of metal. How much more metal is actually required to produce the cans we use? The difference between this and the mathematically optimal can is considerable so there must be some very powerful arguments in favour of the more expensive can.

Notice that the mathematically optimal can has its diameter equal to its height, and therefore presents a squat, dumpy appearance. Furthermore, being only 7.5 cm high, it cannot easily be drunk from when held in an adult hand. Marketing requirements therefore demand that the height of the can must be greater than the width of an adult hand, which in turn leads to a more attractive cross-section.

It is interesting to note that standard 330 ml cans (of height 11.5 cm and radius 3.25 cm) have a diameter to height ratio of 1:1.77 which is not too different from the Golden Ratio 1:1.6. (The Golden Ratio is exhibited by the length to width ratio of many ancient temples and is the length to width ratio of rectangles which artists generally regard as being most aesthetically pleasing.) Given that rectangles whose length to width ratio is close to the golden ratio are deemed 'pleasing to the eye', investigate whether tin can manufacturers would effect significant savings of material if they constructed their 330 ml cans to also satisfy the Golden Ratio condition. Is there a case for recommending that tin can manufacturers change to a Golden Ratio can?

Although this problem has been quite straightforward from a mathematical point of view, some of the issues raised show that some features of some problems can be very subjective and difficult/impossible to quantify.

An Alternative Mathematical Solution

It was mentioned earlier in this example that the constraint equation was easily incorporated algebraically into the expression for the surface area, but that such eliminations were not generally so straightforward. In such cases the constrained optimization problem can be solved by using the method of Lagrange multipliers [4], as is shown next.

Recall that the function to be optimized is

$$S = 2\,\pi rh + 2\,\pi r^2$$

subject to the constraint that the volume is 330 ml, which can be expressed in the form

$$\phi = \pi r^2 h - 330 = 0$$

First construct the associated Lagrangian function

$$S_1 = S + \lambda\phi$$

i.e.

$$S_1 = 2\,\pi rh + 2\,\pi r^2 + \lambda\,(\pi r^2 h - 330)$$

where λ is known as the Lagrange multiplier.

Next set up the three equations

$$\frac{\partial S_1}{\partial r} = 0\ ;\quad \frac{\partial S_1}{\partial h} = 0\ ;\quad \frac{\partial S_1}{\partial \lambda} = 0$$

and solve them for r and h.

DERIVE can be used to determine the partial derivatives by first Authoring the function S_1 with, say, c used to represent λ (suppose it corresponds to expression n) and then in turn Authoring and Simplifying the expressions

DIF (#*n*, *r*), to give, say, expression ℓ

DIF (#*n*, *h*), to give, say, expression *m*

DIF (#*n*, *c*), to give, say, expression *p*

The resulting three expressions are non-linear and so some 'creativity' must be exercised in their solution (we can't just Author them as a set of simultaneous equations).

First soLve expression ℓ for h giving

$$h = -\frac{2r}{cr+1}$$

and again soLve expression m for c giving

$$c = -\frac{2}{r}$$

(Notice that the physically unacceptable solution of $r = 0$ is not displayed.)

Now use Manage and Substitute to substitute c into the above expression for h and Simplify to obtain

$$h = 2r$$

Finally use Manage and Substitute to substitute for h into expression p, Simplify, soLve for r and approXimate to give

$r = 3.744\,93$ (cm) and hence $h = 7.489\,86$ (cm)

as the only physically acceptable (real) solution to the problem.

It is left as an exercise for the reader to confirm that this solution does indeed correspond to a minimum value for the area of metal used.

Further Work

Assume that the can is constructed from two circular discs, representing the top and bottom, and a rectangular piece which is formed into the curved surface. An alternative minimization approach which the can manufacturer might adopt would be to determine the dimensions of the rectangle which minimizes the waste when the three components are cut from it, see Fig. 4.5 in which the shading indicates waste. What are the resulting dimensions if this strategy is adopted? Would this can be likely to satisfy the marketing team's requirements?

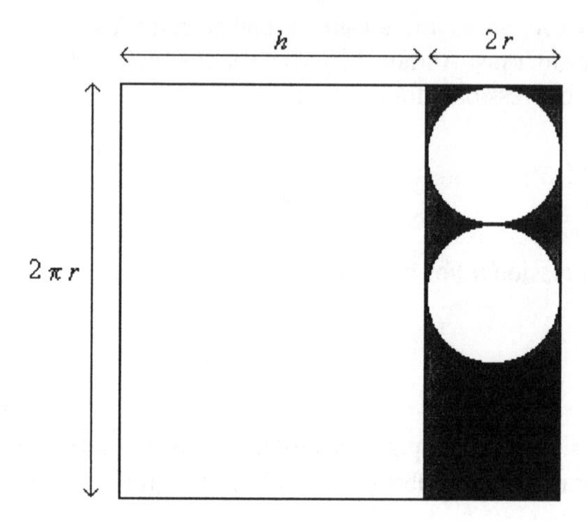

Figure 4.5 Minimizing the wastage in the manufacture of tin cans

4.6 LOCATION OF AN EMERGENCY UNIT

The Problem

A regional health authority has decided to construct an emergency unit adjacent to a section of motorway which runs through the region. For financial reasons it has been decided that the same unit will also have to serve residents of a town situated a few miles from the motorway.

The health authority has decided that the optimal location of the emergency unit on the motorway is the position which minimizes its distance from the town. What advice would you give about the location of the unit?

Setting up a Model

First thoughts are perhaps that this is an optimization problem in co-ordinate geometry. For simplicity, we shall assume that the length of motorway under consideration is straight and that the roads from the town to the ends of the section of motorway are also straight (their curved nature could be taken into account in subsequent refinements). Finally, assume that the new road from the town to the eventual position of the emergency unit is also straight. The assumption that the motorway is straight removes the need (at least initially) to consider the distribution of accidents along it (although this could be built in as a later revision). The problem can then be represented by Fig. 4.6, in which the town is located at the origin and A and B denote the two ends of the section of motorway for which the health authority is responsible.

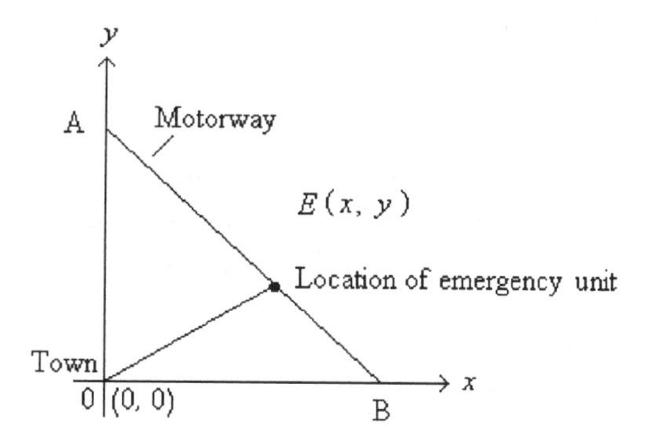

Figure 4.6 Optimal location of an emergency unit

Formulate a Mathematical Model

Mathematically the problem is to minimize the distance *OE*, represented by $\sqrt{(x^2 + y^2)}$, subject to the constraint that $E(x,y)$ must lie on the line AB. For illustration purposes suppose that A and B are situated 12 km and 8 km, respectively, from the town so that the equation of the straight line AB is $y = -3x/2 + 12$. The problem may thus be expressed as

$$\text{minimize } d = \sqrt{(x^2 + y^2)} \text{ subject to the constraint } y = -3x/2 + 12$$

Since minimising d is equivalent to minimising d^2, which will mean we avoid the square root, then we shall pose the problem in the form

$$\text{minimize } d_1 = x^2 + y^2 \text{ subject to the constraint } y = -3x/2 + 12$$

Solution of the Mathematical Model

For constrained optimization problems such as this we can use the method of Lagrange multipliers. We begin by forming the Lagrangian function

$$D = x^2 + y^2 + \lambda\,(y + 3x/2 - 12)$$

where λ is known as a Lagrange multiplier.

Next we determine the three first-order partial derivatives $\partial D / \partial x$, $\partial D / \partial y$ and $\partial D / \partial \lambda$ and set each to zero. The partial derivatives can be obtained easily using DERIVE by first Authoring the expression for D (suppose it is expression k) and then Authoring and Simplifying each of the following expressions

DIF (#k,x), suppose this is expression ℓ
DIF (#k,y), suppose this is expression m
DIF (#k,c), suppose this is expression n

(This last expression corresponds to $\partial D/\partial c$, λ having being substituted with c as λ is not one of the Greek letters available in DERIVE.) Each of these derivatives is linear in x, y and c and so the simultaneous system of algebraic equations

$$\frac{\partial D}{\partial x} = 0 \;, \quad \frac{\partial D}{\partial y} = 0 \;, \quad \frac{\partial D}{\partial c} = 0$$

can readily be solved for x,y (and c) by Authoring the expression

[#ℓ, #m , #n]

soLving it and approXimating the result to give

$$x = 5.538 \,, \qquad y = 3.692$$

The minimum distance from the town is thus approximately 6.7 km.

It is left as an exercise for the reader to confirm that the above values of x and y correspond to a minimum distance OE by checking that

$$\left(\frac{\partial^2 D}{\partial x^2}\right)\left(\frac{\partial^2 D}{\partial y^2}\right) - \left(\frac{\partial^2 D}{\partial x \partial y}\right)^2 > 0 \quad \text{and} \quad \frac{\partial^2 D}{\partial x^2} > 0$$

at the critical point $x = 5.538$, $y = 3.692$, $c = -7.385$.

Interpretation

On the basis of the health authority's declared location policy the distance from the emergency unit to the town will be minimized by locating it at the point (5.54, 3.69), see Fig. 4.6, in which case it will be a distance of 6.7 km from the town.

Further Work

1. A road network in and around the town would obviously already exist. It would be probable that access to the emergency unit from the town would use as much of the existing road network as possible.

   Inclusion of this feature would require knowledge of the geometry of this road network and would lead to further constraints. Investigate for a town and motorway section familiar to you.

2. How could the model be modified to account for bends in the roads and the motorway?
3. The health authority's records might show that accidents on the section AB of motorway were not distributed uniformly. How could you modify the model to account for this?

4.7 OPTIMAL TIMBER CUTTING

Introduction

After trees have been felled in a forest and stripped of their branches they are then delivered to saw mills where they are often cut into beams of rectangular cross-section, the cut being along the length of the log. It is obviously in the interests of the sawmill's profitability to minimize the wastage of the off cuts. What advice could you give the manager about the dimensions of the beam of rectangular cross-section which will minimize the waste when cut from the log?

Setting up a Model

It seems reasonable to assume that the logs are of circular cross-section. Fig. 4.7 shows such a circular cross-section with a rectangle set inside it, the beam that this represents being produced from the log with two horizontal cuts (top and bottom) and two vertical cuts (left and right). The problem reduces to finding the width and height of the rectangle such that its area is maximized while at the same time ensuring that it 'fits' inside the circle. The four sections each having a curved surface represent the waste (of which more later). Symmetry arguments should suggest that the rectangle will in fact be a square – we have left the solution in terms of a rectangle to preserve generality for the purposes of illustration. In fact, if we were solving the problem for ourselves we would exploit the symmetry we had identified (this being good modelling practice).

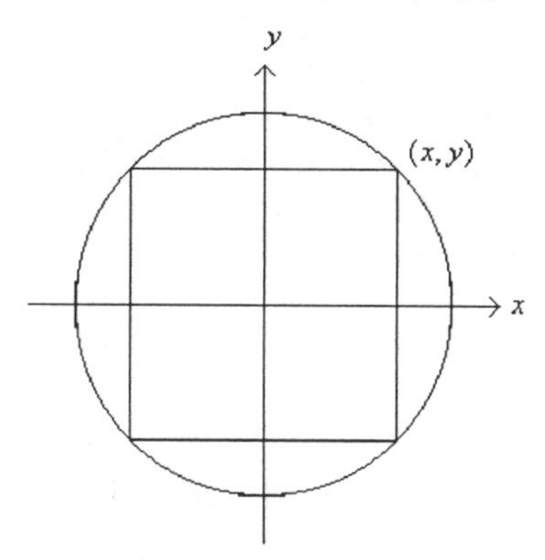

Figure 4.7 Rectangular beam cut from a cylindrical log

The variables on which this problem depends are essentially

- the width of the rectangle ($2x$, see below),
- the height of the rectangle ($2y$, see below),
- the radius of the assumed circular cross-section of the log (r).

To assist in the development of the model it is convenient to introduce a rectangular co-ordinate system (with origin at the centre of the log) and relative to which the co-ordinates of one corner of the rectangle are (x,y), see Fig. 4.7. You may wonder why we haven't suggested the use of polar co-ordinates (prompted by the circular cross-section). While this would make the equation of the circle particularly simple the description of the rectangle would be very difficult (and it is really the rectangle which is of interest).

Formulate a Mathematical Model

The cross-sectional area of the rectangular beam = $(2x)(2y)$ = $4xy$, and this is to be maximized subject to the constraint that

$$x^2 + y^2 = r^2$$

For the sake of argument, suppose the diameter of the log is 1 m, then the mathematical problem to be solved is

$$\text{maximize } 4xy \text{ subject to the constraint } x^2 + y^2 = 0.25$$

This is clearly a constrained optimization problem and we shall solve it using the Lagrange multiplier approach. First, we construct the Lagrangian function

$$A = 4xy + c\,(x^2 + y^2 - 0.25)$$

where c denotes the Lagrange multiplier.

Use the Calculus facility to establish the system of equations from which the optimal result will be obtained as expressions 4, 6 and 7 in Fig. 4.8. Since the equations are non-linear, then we must exercise some interaction with DERIVE to effect their solution. For instance, soLving expressions 4 and 6 for y reveals that

$$-\frac{cx}{2} = -\frac{2x}{c}$$

from which either $x = 0$ (which we can reject as it implies no beam is cut at all!) or $c = \pm\, 2$.

```
1:    "Optimal timber cutting"

                      2     2
2:    4 x y + c (x   + y   - 0.25)

      d                2     2
3:    — (4 x y + c (x   + y   - 0.25))
      dx

4:    2 (c x + 2 y)

5:    "Optimal solution given by solution of #4 and"

6:    4 x + 2 c y

       2     2
7:    x   + y   - 0.25
```

```
COMMAND: Author Build Calculus Declare Expand Factor Help Jump soLve Manage
         Options Plot Quit Remove Simplify Transfer moVe Window approX
Enter option
User                                    Free:100%             Derive Algebra
```

Figure 4.8 Formulation of the timber cutting optimization problem using DERIVE

Next, use Manage and Substitute on expression 4 and 6 to substitute either $c = +2$ or $c = -2$ and then soLve for y to establish that

$$y = \pm x$$

Finally use Manage and Substitute on expression 7 to substitute for y (as either $+x$ or $-x$) and soLve for x to give

$$x = \pm \sqrt{2}/4 \qquad \text{so that } y = \pm \sqrt{2}/4$$

Interpretation

The beam of maximum cross-section which can be cut from a circular log of radius 1 m has a square cross section of side $1/\sqrt{2}$m. The cross-sectional area of the beam is then 0.5 m^2, which indicates a substantial wastage as the cross-sectional area of the log is π m^2.

Further Work

Fig. 4.9 shows a possible way of reducing the wastage by cutting four secondary beams from the initial waste.

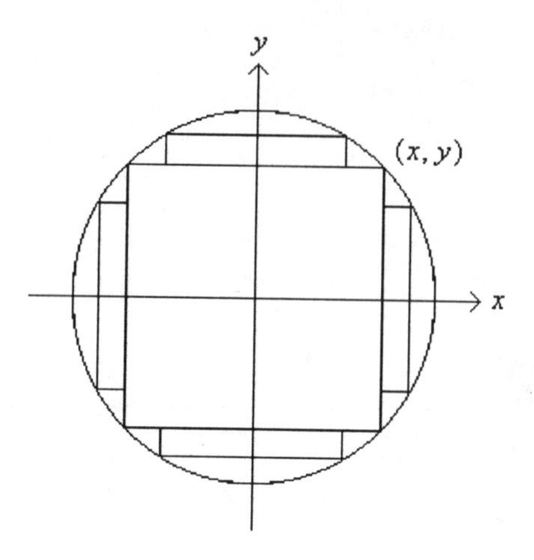

Figure 4.9 Possible further cuts to reduce wastage

What should be their optimal dimensions?

What is the wastage now? Do you consider this to be commercially acceptable? If so what might the waste be used for? If not, what could be done next to further minimize the wastage?

References

[1] Galvin W., 'Machine scheduling models with applications to the loading of bulk carriers in tidal ports', *IMA Bulletin, 21* (1/2), January/February, 21–25 (1988).

[2] Fryer M. J., *An Introduction to Linear Programming and Matrix Game Theory*, Arnold (1987).

[3] Townend M. S., *Mathematics in Sport*, Ellis Horwood, Chichester (1984).

[4] Stroud K.A., *Further Engineering Mathematics*, 2nd edition, Macmillan (1994).

5

Statistical Modelling

5.1 INTRODUCTION

Many problems require the modeller to attempt to formulate a prediction of behaviour based, perhaps, on the use of the ideas of probability, or a prediction based upon a collection of empirical data. The study of 'statistics' is a significant component of mathematics, engineering and science undergraduate degree programmes, and statistical modelling has been applied to many areas of these disciplines. Here we describe four case studies that involve four aspects of a traditional statistics course, namely, forecasting trends using least squares, probability distribution models, Markov chains and regression. The intention is to emphasize the modelling aspects of the case study rather than present a comprehensive exposition of the underlying statistical theory. This philosophy is reflected in the use of DERIVE, which contains simple statistical functions and probability distributions but does not have the complete range of utilities found in commercial statistical software packages.

5.2 TRENDS IN OLYMPIC PERFORMANCE

Forecasting Trends

The ability to forecast stock market prices, football results, political results, and so on, accurately, has led to much research into methods of forecasting and the determination of trends. Clearly, an accurate procedure for predicting the future behaviour of stock market prices would enable people to make their fortunes, although as a modelling aside it is worth noting that if too many people possessed such forecasting skills there would be no fortune to be made!

Many forecasting techniques have been developed and in the following case study we shall examine only one, with others being left for you to investigate.

The Problem

Every four years the Olympic Games becomes the focus of attention of the world's sporting élite. All the participants have honed their preparations for the heats and finals of their specific event(s) and two thoughts uppermost in their minds are whether their performance will gain them a medal and, if so, whether it will surpass previous Olympic performances. (We have deliberately used the Olympic Games as a vehicle with which to illustrate a forecasting technique as this is considered to remove subject variability with regard to motivation and competitive conditions.)

It is required to develop a model which can be used to investigate any trends in Olympic winning performances.

Setting up a Model

First we consider some of the features on which Olympic performance is likely to depend:

- health of the athlete,
- number of athletes,
- effectiveness of the coaches,
- climatic conditions at the time,
- political conditions at the time (boycott),
- geographic features of the venue,
- previous winning performance,
- psychological profile of the athlete, ...

Next we turn our attention to establishing the assumptions from which a model will be developed. We shall assume the following for any specific events considered:

- There is a sufficient number of healthy athletes.
- Under the guidance of their coaches each individual athlete has reached his or her optimal level of fitness and performance.
- Climatic conditions are ignored (although they might have a bearing on endurance events).
- The Games have not been boycotted by anyone.
- Geographic conditions are ignored.
- Psychological factors are ignored in the belief that athletes who reach an Olympic final have such factors under control. They are also difficult to quantify in our model!
- Performances in the events of an Olympiad are related to performances in the corresponding events in the previous Olympiad.

Formulate a Mathematical Model

The final assumption made above suggests that an appropriate initial model may well be established by developing a simple linear model of the form

$$\begin{pmatrix} \textit{Winning} \\ \textit{performance} \\ \textit{in current} \\ \textit{Olympiad} \end{pmatrix} = \textit{constant} \times \begin{pmatrix} \textit{winning} \\ \textit{performance} \\ \textit{in previous} \\ \textit{Olympiad} \end{pmatrix} + \textit{error term} \tag{5.1}$$

where the error term is a measure of the inaccuracy of the estimate and may vary from one Olympiad to another.

If we define the current Olympiad to be the nth competition and introduce the following variables

winning performance in current Olympiad	W_n
'constant'	c
error term (dependent on the counter n)	e_n

then our model can be expressed mathematically as

$$W_n = c\, W_{n-1} + e_n \qquad n = 2, 3, \ldots \tag{5.2}$$

where W_1 denotes the winning performance in the first Olympiad considered.

Solution of the Mathematical Model

The 'best' model will be the one which minimizes the error term e_n. One way of achieving this is to use the method of least squares. In this the optimal value of the constant c is found by minimizing the sum of the squares of the error terms i.e. by minimizing

$$\sum_{n=2}^{m} e_n^2 = \sum_{n=2}^{m} (W_n - c\, W_{n-1})^2 \tag{5.3}$$

where there have been m previous Olympiads. A similar model has been developed and reported by Stefani [1].

It is a simple exercise to establish that, using the least-squares criterion of 'best', the optimal value of c is given by

$$c = \sum_{n=2}^{m} W_n\, W_{n-1} \Big/ \sum_{n=2}^{m} W_{n-1}^2 \tag{5.4}$$

Note: (5.4) is obtained by differentiating (5.3) with respect to c, setting the result equal to zero and soLving for c. You can use DERIVE's facilities to confirm the result.

Interpretation

Olympic events fall into the following two broad categories:

- those in which improvements correspond to a reduction in the measured parameter (e.g. time for track events),
- those in which improvements correspond to an increase in the measured parameter (e.g. the weight lifted in a weightlifting competition).

The former category therefore corresponds to $c < 1$, while the latter category corresponds to $c > 1$.

Stefani [2] has developed a convenient index of improvement which is called the percent improvement per Olympiad (% *I/O*) and is defined as

$$\% I / O = \begin{cases} 100(c-1) & \text{for measured events} \\ 100(1-c) & \text{for timed events} \end{cases}$$

Based on this index Stefani reports a number of interesting trends by examining Olympic results over both the long term for men (many previous Olympiads) and the short term per event (men and women). The results are presented in Figs. 5.1, 5.2, which are reproduced from [2].

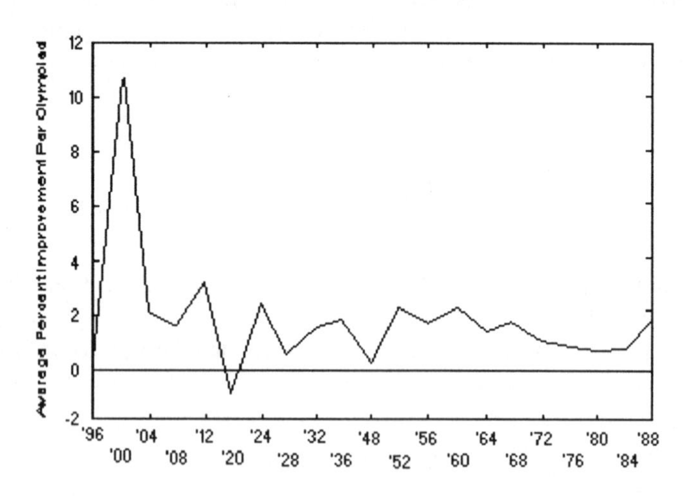

Figure 5.1 Percentage improvement for each Olympiad averaged for men's athletics

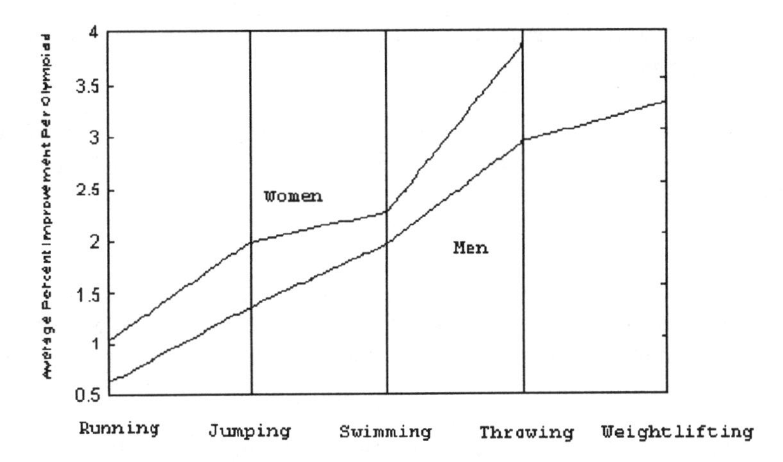

Figure 5.2 Percentage improvement per Olympiad 1952–1988 average by type of event

Reproduced with permission from 'Applying least squares to team sports and Olympic winning performances' by Stefani R.T., published in *Proceedings of Mathematics and Computers in Sport Conference, Bond University, Australia*, July 1992.

Use your modelling expertise to convince yourself that the following trends represent plausible conclusions

- The rate of improvement for women is greater than that for men. You may care to consider why this should be so – consideration of the marathon may be helpful.
- The rate of increase for either sex follows the increase in leverage of the disciplines involved.

The chronological trends, too, are extremely interesting. Observe the following:

- Increased nationalism in the immediate pre-War Olympiads (1912 and 1936) led to large increases in performance.
- There was a marked reduction in performance in the immediate post-War Olympiads (1920 and 1948). Why?
- As the Cold War period (around 1950–1970) saw intense sporting rivalry between the USA and the, then, USSR there was a consequent heightened rate of improvement.
- Reductions in improvement were apparent at the time of various recent boycotts (by African nations in 1976, Western bloc nations in 1980, Eastern bloc in 1984). Why?

- There were increased improvements in 1988 when all nations returned to Olympic competition.

Stefani's conclusions have been presented in detail as they are an excellent example of the 'lateral thinking' in which modellers must be prepared to engage in order to explain their results fully.

Validation

To assist you in assessing the validity of the model developed here we present the results for the winning time in the Olympic final of the men's 100 m race for the period 1948–1992. .

1948	Dillard	(USA)	10.3 s	1972	Borzov	(USSR)	10.14 s
1952	Remigino	(USA)	10.79 s	1976	Crawford	(TRI)	10.06 s
1956	Morrow	(USA)	10.62 s	1980	Wells	(UK)	10.25 s
1960	Hary	(FRG)	10.32 s	1984	Lewis	(USA)	9.99 s
1964	Hayes	(USA)	10.06 s	1988	Lewis	(USA)	9.92 s
1968	Hines	(USA)	9.95 s	1992	Christie	(UK)	9.96 s

Use the results for the period 1948–1988, inclusive, to predict the result for 1992. How does it compare with Christie's time?

Use the results for the period 1948–1964 to predict the 1968 result. Can you suggest reasons for the discrepancy?

Note. You may find it useful to perform the calculations with DERIVE as follows. Author a vector containing the eleven time values

e.g. $W := [\ 10.3,\ 10.79,\ \ldots,\ 9.92\ ]$

and then compute c from equation (5.4) by Authoring

$$c := \text{sum (element } (w,n) \ * \ \text{element } (w,n-1),\ n,\ 2,\ 11)$$
$$/(\text{sum (element } (w,n-1) \ * \ \text{element } (w,n-1),\ n,\ 2,\ 11))$$

Having an approXimate value for c, multiply the 1988 time value by c to predict the 1992 time for Christie. Similar DERIVE expressions can be used when working with subsets of the complete data set.

Comment

It is worthy of note that Stefani's index of improvement gives a simple measure of relative performances between track and field events. Performance in track events is directly proportional to time, whereas for field events the measure is the distance thrown or jumped. Since the range of a projectile is proportional to the square of the release velocity, then meaningful comparisons between track and throwing events are complicated by the fact that

track performance	$\propto$	time
field performance	$\propto$	1/time squared

i.e.

log (event time)	$\propto$	1
log (distance thrown)	$\propto$	−2

Points tables for the decathlon reflect this factor of two difference between track and field, but how much simpler is the measure adopted by Stefani!

Further Work

1. Repeat the estimation of Christie's 1992 time using the full history by utilizing DERIVE's FIT function. Create a matrix

   $$D:=\left[[1,\ 10.3]\ ,\ [2,\ 10.79]\ ,\ \ldots,[11,\ 9.92]\right]$$

   and then Author

   $$\text{FIT}\left(\left[x,\ ax+b\right]\ ,\ D\right)$$ and Simplify

   Use the result to compute the 1992 predicted time (i.e. $x = 12$). Plot both the data set and the fitted line. Does the straight line fit look reasonable?
2. Maybe it would be more appropriate to use a second-order model of the form

   $$W_n = a\,W_{n-1} + b\,W_{n-2} + e_n\,?$$

   Investigate this model using DERIVE's facilities to assist with the algebra of the least-squares calculations and the subsequent model fitting.
3. Select other Olympic events of interest to you and investigate their trends.
4. As mentioned earlier, many other forecasting techniques are available. One such is based on the idea of moving averages. For example, a trend line can be obtained by dividing the data into two, finding the average value of each set of data values and then fitting a straight line to the two averages.

   Investigate this and other techniques for estimating trend lines and apply them to the 100 m results. Use DERIVE to help you perform the calculations. Are your estimations and conclusions significantly changed?

5.3 SHOVE HA'PENNY

Introduction

In order to generate some funds for their youth club, a group of members decide to run a shove ha'penny stall at a church fête. The game involves rolling a coin down an inclined groove onto a horizontal table on which is drawn a set of parallel lines. If the coin stops between the lines it is a 'winner' but if it stops across or touching a line it is a 'loser'.

Participants roll four coins (the members have suggested using 2p pieces) each time they play. If all four coins are 'winners' they receive a prize of 50p (for their 8p stake), if three are 'winners' they receive a refund of their stake money; otherwise the club makes 8p profit.

The Problem

The youth club leader knows that you are a mathematician and approaches you for advice with regard to the likely financial benefit to the club, an assessment as to whether the game is sufficiently enticing and an idea of a reasonable separation of the parallel lines.

Develop a model which addresses these issues and decide what advice you would give to the youth club leader.

Setting up a Model

The following would be some relevant features of the problem:

- the diameter of the coin used,
- the separation of the parallel lines,
- the number of coins rolled in one complete play,
- the attraction of the prize , ...

Reasonable assumptions from which to begin the development of a model might be as follows:

- The result of rolling each coin is independent of any previous results.
- The four coins used are all of the same diameter (25 mm for the suggested 2p piece).
- Each complete play consists of rolling the same number of coins (four in the proposed game).

- The parallel lines are equally spaced (d mm) and at right angles to the direction of approach of the coin down the inclined groove.

Formulate a Mathematical Model

The modelling assumptions suggest that the binomial distribution is applicable to the solution of this problem (after all, each coin is either a 'winner' or a 'loser'). For completeness, and safety's sake (?!), we give a brief reminder:

$$\Pr\{\, k \text{ successes in } n \text{ trials} \,\} = {}^{n}C_{k}\; p^{k}\,(1-p)^{n-k}$$

where p denotes the probability of success, and

$${}^{n}C_{k} = \frac{n!}{(n-k)!\,k!}$$

The probability of any one coin being successful (from the point of view of the player) depends on the separation of the lines. Suppose the separation is d (mm), $d > 25$, then for the two pence piece proposed for the game there is a margin of width (d–25) mm in which the coin can land without touching/crossing a line. Hence we can say that

$$\Pr\{\text{ coin is successful }\} = (d-25)\,/\,d$$

In order to develop the model numerically, a line separation of 30 mm is proposed, although this is obviously one feature which we can subsequently vary when investigating the sensitivity of the solution. Thus

$$\Pr\{\text{ coin is successful }\} = 5/30 = 0.1667$$

The task is then to develop a model based on the binomial distribution and which looks at all the possible outcomes of one complete play and subsequently to interpret the resulting statistics.

Solution of the Mathematical Model

The following table shows the various possible outcomes of one complete play and the associated gain to the youth club. The statistical component of the calculations was performed using the BINOMIAL_DENSITY function within the DERIVE utility file PROBABIL.MTH.

Number of winning coins	4	3	2	1	0
Probability	0.000 772	0.015 441	0.115 778	0.385 833	0.482 176
Gain to youth club	−42p	0p	8p	8p	8p

The expected income per participant (calculated from the expression for the statistically expected value of a variable) is then given by

$$-42\times 0.000\,772+8\times 0.115\,778+8\times 0.385\,833+8\times 0.482\,176=7.837\,872\simeq 7.8\text{p}$$

Interpretation

The expected income per participant shows that from the organizer's point of view the game is quite rewarding. But what do things look like from the player's point of view? The table of probabilities shows that about 98% of the players come away with nothing. Hence although the organizers might feel that there is a small fortune to be made they are in fact unlikely to find that the game has many players! (Modelling reminder – in this sort of problem don't forget to look at the solution from everyone's point of view!)

Development of a Revised Model

It is obviously important to try to make the game more attractive to potential players. One way to do this would be to relax the rules sufficiently to increase its attraction to a player while still ensuring a reasonable income for the youth club.

For example, instead of refunding the stake money to players who achieve three 'winners' in a complete play we might instead offer a further free play (using all four coins) to players who achieve two or three 'winners'. This will clearly have an effect on the probability distribution – it would be possible for a player to achieve two or three 'winners' time after time. How are the results affected by this change in the rules?

Solution of the Revised Model

It is helpful to set out the various play continuing possibilities in terms of successful coins (S) and failures (F) for the revised model.

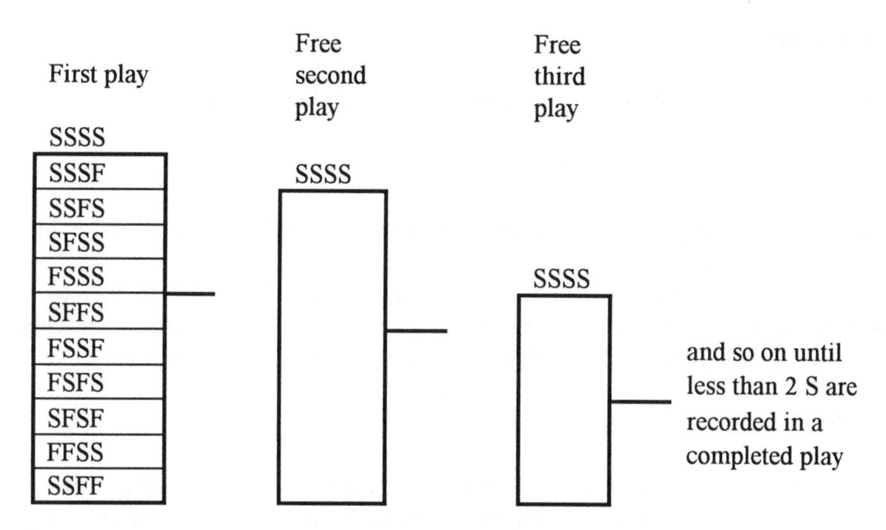

It is evident that the youth club makes a loss of 42p if the first play, or the second play or the third play or ... results in four successes; otherwise it makes a profit of 8p.

Using the rules of elementary probability:

Pr {4 successes with 4 coins in the first play} $= 0.000\,772$

Pr { 2 or 3 successes in the first play and 4 successes with 4 coins in the second play} $= (0.115\,778 + 0.015\,441) \times 0.000\,772$

$= 0.131\,219 \times 0.000\,772$

Pr { 2 or 3 successes in the first and second plays and 4 successes with 4 coins in the third play} $= 0.131\,219^2 \times 0.000\,772$ ...

Hence the total probability that a player ultimately wins 50p is

$$0.000\,772\,(1 + 0.131\,219 + 0.131\,219^2 + \ldots)$$

The parenthesised term is recognized as a geometric series whose sum to infinity can be obtained using DERIVE's SUM facility within the Calculus command to give the above probability as

$$0.000\,772 \times 1.151\,038 = 0.000\,889$$

Interpretation

The revised model proposed above leads to an expected profit per player of

$$-42 \times 0.000\,889 + 8(1 - 0.000\,889) = 7.955\,55 \quad \simeq 7.96\text{p}$$

The chances of a player winning are enhanced to about 13% (i.e. 0.000 772 + 0.015 441 + 0.115 778 × 100%).

When played according to the revised rules the game is therefore more attractive to both parties, although it might be argued that a 13% chance of winning is not very high, perhaps not high enough to make the game compelling.

Further Work

1. Investigate the original game and the proposed revision for different values of the line spacing d mm. Can the game be made sufficiently attractive to players solely by changing the spacing? Propose your own revisions and investigate their solutions. Investigate the effect of changing the number of coins used in one complete play.
2. Another consideration in this game is the time it takes to earn funds. For example, the youth club would probably be happier to gain less income per 'go' by a participant if they could ensure more 'goes' (i.e. fewer free plays) in a given time period. Suppose that it takes two minutes on average to roll the four coins, and the church fête lasts for two hours. How will your model(s) be revised to optimize earnings?
3. Besides the actual playing of the game, one way to see if the model is realistic is to simulate the game on a computer using a random number generator. As an example, if one considers the centre of the coin, then as it takes all possible positions between two lines, its distance from one line varies between 0 and 30 (mm) inclusive. The coin is a winner whenever this distance lies between 12.5 and 17.5 mm.

   Authoring the DERIVE expression 30 * RANDOM(1) generates a random number in the interval [0, 30]. Use this to generate a number of plays and count the winners. Does the expected income from your simulation agree with the predicted income in the long run?
4. A variation of this game is to divide the table into squares and label each with a number. If the coin lands within a square the player is paid the number of pence with which the square is labelled. Investigate how the squares might be labelled numerically in order to produce a game that is attractive to both the players and the organizers.

5.4 WINNING AT BADMINTON

The Problem

Readers may be familiar with the game of badminton, either by playing at club level or higher or by watching the game on television. Badminton can be played by two opposing players in a game of singles or by opposing pairs in a doubles match. When played at club level in particular, the rules of the game allow for different ways of ending a set, which can be summarized as the first player to score 15 points wins, BUT, if the score is tied at 13–13, the player who reaches 13 first can decide either to play through to 15 OR set the score to 0–0 and play first to 5 points; if the score is 14–14, the player who reaches 14 first can decide whether to play through to 15 OR set the score to 0–0 and play first to 3 points.

Points can only be won by a player when he/she is serving. Serving starts from the right-hand court and the server serves diagonally to the opposition, who receives in their right-hand court. The serve alternates from the right-hand to the left-hand side, and so on, as the score changes. The rule that most players remember is 'if your score is an odd number then serve from the left'. A match typically consists of the best of three sets.

Can we devise a model of a player's performance that will enable that player to select a strategy for ending a set that will improve the chances of winning it?

Setting up a Model

A possible list of features that could be included in such a model is the following:

- the likelihood of a player scoring a point (this can only be done when serving,remember!),
- the likelihood of a player winning (but not scoring a point) when the opponent is serving,
- which player serves first,
- the likelihood of a 13–13 or 14–14 score line,
- which side the serve is from (and the effect this may have on the first two points in this features list),
- physical and environmental factors, e.g. which end is the player at, how tired a player becomes as the match progresses (and hence the first two points on this list are affected again), etc.

Initially, we shall assume that the likelihood of winning a serve is not influenced by the side of service. The last feature will also be disregarded to allow for the development of a simple model. We shall also consider a singles match.

Formulate a Mathematical Model

Let us suppose that there are two players, player A and player B, and that the probabilities are defined as:

p_A : the probability that A wins a rally when serving

q_A : the probability that A wins a rally when receiving

Then from the laws of probability

$p_B = 1 - q_A$: the probability that B wins a rally when serving

$q_B = 1 - p_A$: the probability that B wins a rally when receiving

Consider the possibilities for a given point to be scored by A:

A could serve, win the rally and score the point. A could serve and lose the rally. B could then serve and lose the rally. A could then serve and win the rally, etc.

Similar considerations arise for B scoring a point with A serving first. Thus we shall further define probabilities:

f_A : the probability that A scores a point when serving first
g_A : the probability that A scores a point when receiving first

and hence

$f_B = 1 - g_A$: the probability that B scores a point when serving first
$g_B = 1 - f_A$: the probability that B scores a point when receiving first

Then the consideration of all possible rallies gives

$$f_A = p_A + q_B\, q_A\, p_A + q_B\, q_A\, q_B\, q_A\, p_A + \ldots \tag{5.5}$$

and $$g_A = q_A\, p_A + q_A\, q_B\, q_A\, p_A + q_A\, q_B\, q_A\, q_B\, q_A\, p_A + \ldots \tag{5.6}$$

Now as the set progresses, for a given scoring point, the players are in one of two 'states', either serving (S) or receiving (R). When a point has been scored, then a state transition can occur. In our notation we have:

A : S → S	with probability f_A
A : S → R	with probability $g_B = 1-f_A$
A : R → S	with probability g_A
A : R → R	with probability $f_B = 1-g_A$

with similar statements for player B.

This is an example of a two-state Markov chain. (For further details see reference [3], for example.) Such a chain is characterized by a transition matrix T with entries t_{ij}, where t_{ij} is the probability of change of state from j to i.

If we set i to be the state 'serving' (S), and j to be the state 'receiving' (R) then the transition matrix becomes

$$T = \begin{bmatrix} f_A & g_A \\ 1-f_A & 1-g_A \end{bmatrix} \tag{5.7}$$

As the points are won and the states change, the matrix T^n, where n is the number of points played, gives an indication of the likelihood of the serving state (and hence who is winning) as the set and match progress.

Solution of the Mathematical Model

Numerically, the outcomes hinge upon the values of p_A and q_A. Before any values are used, however, we can make some mathematical observations about equations (5.5), (5.6) and (5.7).

The infinite series (5.5) and (5.6) can be reorganized as a geometric series of the form

$$a + ar + ar^2 + ar^3 + \dots$$

which has a 'sum to infinity'

$$S_\infty = \frac{a}{1-r}$$

Therefore, (5.5) and (5.6) become

$$f_A = \frac{p_A}{1-q_A q_B} \quad \text{and} \quad g_A = \frac{q_A p_A}{1-q_B q_A} = q_A f_A \tag{5.8}$$

The matrix T is a matrix with non-negative elements, all of which lie between 0 and 1 inclusive, and each column sums to unity. Such a matrix is termed a STOCHASTIC or MARKOV matrix (see also the forest management study in Chapter 3). Such matrices have the following properties:

(i) One of the eigenvalues must equal 1
(ii) All eigenvalues, λ_i, satisfy $|\lambda_i| \le 1$
(iii) If T is stochastic, then so is T^n for some integer n
(iv) T^n tends to a limit as $n \to \infty$, independent of the initial state, if and only if one of the eigenvalues equals unity and all the others have magnitude less than unity.

This final property is particularly useful when estimating who will win 'in the long run'.

As a specific example, suppose that A considers him(her)self to be stronger than B and assigns probability values $p_A = {}^3/_4$ and $q_A = {}^2/_3$ to him(her)self. Then (5.8) gives

$$f_A = {}^9/_{10} \qquad \text{and} \qquad g_A = {}^3/_5$$

Use DERIVE's function EIGENVALUES to compute the eigenvalues of T in this case by Authoring

$$\text{EIGENVALUES}\left(\left[\left[{}^9/_{10},\ {}^3/_5\right],\ \left[{}^1/_{10},\ {}^2/_5\right]\right]\right)$$

and Simplify. You should observe that the computed eigenvalues satisfy the above properties and hence T^n will tend to a limit as $n \to \infty$. Use DERIVE to observe how T^n varies as n increases.

Author the expression

$$\left[\left[{}^9/_{10},\ {}^3/_5\right],\left[{}^1/_{10},\ {}^2/_5\right]\right] \wedge n$$

where n is to be set to some integer value. How does T^n vary as n takes values 5, 10, 15, 20, say? After about 20 points you should observe that T^n settles to a steady value that predicts that A will win 85.7% of the points in the long run.

Interpretation

Clearly, a given player is unlikely to be able to quantify a value for p_A and q_A. In many cases, especially when playing in matches where players are ranked by past performances, it is perhaps not unreasonable to consider $p_A = {}^1/_2$, i.e. equally likely to win or lose a rally whether serving or not. What does the model predict in this case?

Firstly, if $p_A = q_A = {}^1/_2$ then (5.8) gives $f_A = {}^2/_3$, $g_A = {}^1/_3$ and this suggests immediately that serving first is an advantage (why?).

Secondly, suppose the score has reached 14–14 with player A having reached 14 first and player B currently serving. What strategy for A does the model suggest? If player A chooses to play to 15, then the probability (s)he wins is $g_A = {}^1/_3$. If (s)he chooses to set 'to first to 3 points', then the possible winning sequences are (player A first) :

{1–0, 2–0, 3–0}	{1–0, 1–1, 2–1, 2–2, 3–2}
{0–1, 1–1, 2–1, 3–1}	{1–0, 1–1, 1–2, 2–2, 3–2}
{1–0, 1–1, 2–1, 3–1}	{0–1, 0–2, 1–2, 2–2, 3–2}
{1–0, 2–0, 2–1, 3–1}	{0–1, 1–1, 2–1, 2–2, 3–2}
{1–0, 2–0, 2–1, 2–2, 3–2}	{0–1, 1–1, 1–2, 2–2, 3–2}

The probability of winning is the sum of the probabilities of these events and the evaluation shows that this probability is greater than g_A (but less than ½ ??). So at 14–14, the optimal advice for players of equal ability is to set to 3 points.

How likely is a score of 15–0? Again, if A and B are of equal ability and $p_A = q_A = {}^1/_2$, then the probability of A winning 15–0 is:

(i) $({}^2/_3)^{15}$ when A serves first,
(ii) $({}^1/_3)\,({}^2/_3)^{14}$ when B serves first,

i.e. there is a likelihood of A serving first and winning 15–0 in about two sets in every 1000.

What is the probability that A wins a set if (s)he starts serving? To calculate this it is necessary to add the probabilities of winning by the various possible scores, i.e.

$$\text{prob}\,\{15–0\} \;+\; \text{prob}\,\{15–1\} \;+\; \ldots$$

(remember to include the setting possibilities).

The calculations are left to the reader (and DERIVE), but it is possible to show that serving first is more likely to lead to a set win (but is it a significant advantage?).

Validation

Is this model reasonable? There are several ways in which it could be validated by comparing its predictions with past results. For example, international match scores could be examined to see if the prediction of two sets in every 1000 won by 15–0 has in fact happened. One obvious criticism is that as players tire, p_A and q_A might vary in a set or match, but this is difficult to quantify numerically *a priori*.

Further Work

1. For the model presented answer the following:

   (a) For $p_A = q_A = {}^1/_2$ and given that the score is 13–13 and player A reached 13 first, confirm that the optimal strategy for A is to set for 5 points.

   (b) Is there a value or set of values for p_A for which A's optimal strategy at 14–14 is to play to 15?

   (c) Supposing A serves first and $p_A = q_A = {}^1/_2$, calculate the most likely set score and hence the most likely match score. (These results would provide a further means of validating the model by comparing with international match statistics.)

   (d) A player decides to concentrate hard on his/her serve and to 'go for broke' when receiving serve against a player of comparable ability, i.e. $p_A = {}^2/_3$, $q_A = {}^1/_3$ is thought more appropriate than $p_A = q_A = {}^1/_2$. Does this strategy give a better chance of winning?

   (e) For $p_A = q_A = {}^1/_2$, what is the probability that A will score the next point 'in the long-run'? (Write down the transition matrix T and compute T^n for increasing n.)

2. The case study for badminton presented here parallels the model developed by Alexander, McClements and Simmons in [4] for the game of squash. In squash, as in badminton, you can only score when serving and a set is the first to 9. If the score is 8–8, then the player to reach 8 first can select to play to 9 or to set to two points. Develop a model similar to the above and determine player A's best strategy at 8–8, (see [4] for confirmation of your results).

3. The game of volleyball scores similarly to badminton. Investigate the rules of scoring and develop a model appropriate for this game.

4. Tennis is a game that does score differently from badminton in that a player can score a point and win a game when serving or receiving. The tournament organizer of the Wimbledon Tennis Championships would find it useful to know the likelihood of the men's singles final lasting three sets or four sets or five sets. Investigate the rules of scoring in men's singles tennis and formulate a model that will help the tournament organizer.

5.5 BIOLOGICAL DATA MODELLING

Introduction

Many statistical applications involve modelling the relationships among sets of variables, especially when evidence for a relationship is based on measurements of the variables. In some cases there is an *exact* relationship between the variables, and the model that describes this relationship is termed a DETERMINISTIC model. Examples of such models are the relationship between applied force and acceleration of a body (Newton's laws of motion) and the relationship between voltage and current across a resistance (Ohm's law). Of course, any experiment to obtain measured data to confirm these laws will generate random error because of measurement inaccuracies.

Often, there is no known underlying law to suggest a deterministic relationship, and a more appropriate model is one that tries to account for random error due to data measurement inaccuracies or some other random phenomenon. Such a model is termed a PROBABILISTIC model.

In this case study we discuss one method of generating a probabilistic model called regression analysis.

The Problem

A biologist colleague of the authors has been investigating the growth characteristics of a certain type of lizard found in the Canary Isles and has gathered the following experimental data:

Lizard Number	1	2	3	4	5	6	7	8	9	10	11	12	13
Body size x (trunk length in mm)	37.9	39.6	33.0	32.4	31.2	26.8	50.7	48.5	53.4	50.3	52.9	55.5	48.2
Distance from base of neck to tip of snout (mm) y	19.6	21.3	18.2	18.0	17.3	15.5	25.4	22.6	26.3	24.0	23.0	26.7	23.8

Lizard Number	14	15	16	17	18	19	20	21	22	23	24	25	26
Body size x (trunk length in mm)	52.2	34.8	34.4	33.4	29.8	29.9	43.5	48.8	53.5	53.5	49.6	37.5	37.0
Distance from base of neck to tip of snout (mm) y	23.9	18.2	17.7	16.8	15.7	17.0	23.0	23.8	25.9	26.6	25.9	20.0	20.2

The corresponding data values are linked, i.e. the first lizard measured has a body length of 37.9 mm and a neck-to-snout measurement of 19.6 mm, etc., twenty-six lizards being measured in total.

Can the lizard data be used to formulate a model that can be used to predict neck-to-snout length for a given body length?

Setting up a Model

Perhaps the first action to take with any data is to plot the values and see what results. This plot is usually referred to as a scattergram and is easily obtained using DERIVE by

(i) Authoring a vector **d** that contains the 26 data pairs as elements.

d : = [[37.9, 19.6] , [39.6, 21.3] , ...,[37.0, 20.2]]

(ii) Plotting the vector **d** using appropriate scales. (Centre and Zoom as necessary.)

The plot obtained is shown in Fig. 5.3, where the horizontal axis corresponds to the body size data and the vertical axis to neck-to-snout data. We shall represent body size data by x and neck-to-snout data by y from now on.

What does the scattergram tell us? For the range of values plotted, there is clearly no simple deterministic model that can relate x and y exactly, but it looks as if a model with a straight line component and an error component may be suitable. Hence we propose a model of the form

$$y = \beta_0 + \beta_1 x + \varepsilon \tag{5.9}$$

where β_0 and β_1 are values to be determined using the supplied data values and ε is the random error component.

To develop the statistical analysis we shall make the following assumptions:

- The values of β_0 and β_1 can be determined using a least-squares analysis (as in the forecasting study).
- The error component ε is normally distributed with mean zero and constant variance, and errors associated with different observations are independent.
- The supplied data values are exact, i.e. measurement errors in both x and y are ignored.

Various statistical techniques exist for testing the validity of these assumptions. These are generally beyond this scope of this book, but the interested reader should consult [5], for example.

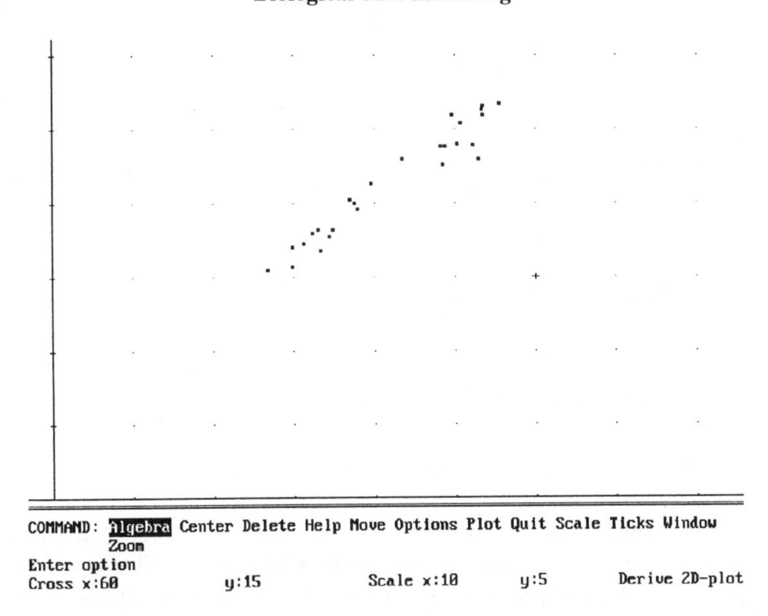

Figure 5.3 Scattergram of lizard data

Formulate a Mathematical Model

Having hypothesized a form for the relationship between x and y in equation (5.9), we now attempt to estimate β_0 and β_1 and the parameters of the distribution of ε. Using the least-squares approach, we choose the estimate

$$\hat{y} = \hat{\beta}_0 + \hat{\beta}_1 x$$

so that the sum of squares of the differences between the observed values y_i and their estimates $\hat{y}_i = \hat{\beta}_0 + \hat{\beta}_1 x_i$ is minimized, i.e.

$$\sum_{i=1}^{n} (y_i - \hat{y}_i)^2 = \sum_{i=1}^{n} \left(y_i - \hat{\beta}_0 - \hat{\beta}_1 x_i\right)^2 \qquad \text{is minimised}$$

where n is the number of data values.

This is achieved by partial differentiation with respect to both $\hat{\beta}_0$ and $\hat{\beta}_1$ and setting the resulting expressions to zero. The results are

$$\hat{\beta}_1 = \frac{SS_{xy}}{SS_{xx}} \quad , \quad \hat{\beta}_0 = \bar{y} - \hat{\beta}_1 \bar{x}$$

where

$$SS_{xy} = \sum_{i=1}^{n}(x_i - \bar{x})(y_i - \bar{y}) = \sum_{i=1}^{n} x_i y_i - \frac{\left(\sum_{i=1}^{n} x_i\right)\left(\sum_{i=1}^{n} y_i\right)}{n}$$

$$SS_{xx} = \sum_{i=1}^{n}(x_i - \bar{x})^2 = \sum_{i=1}^{n} x_i^2 - \frac{\left(\sum_{i=1}^{n} x_i\right)^2}{n}$$

$\bar{y}$ = arithmetic mean of the y data values

$\bar{x}$ = arithmetic mean of the x data values

(the disbelieving reader is again encouraged to reproduce these results using DERIVE).

The above formulae can be computed using DERIVE's utilities in a number of ways, but the authors have found it most convenient to proceed as follows. First declare vectors **x** and **y** containing the respective data values.

(i) Author **x** : = [37.9, 39.6, . . . , 37.0] and
y : = [19.6, 21.3, . . . , 20.2]

(ii) use the statistical function AVERAGE to compute $\bar{x}$, $\bar{y}$ and also $\sum_{i=1}^{n} x_i$, $\sum_{i=1}^{n} y_i$

i.e. $\bar{x}$ can be found by Authoring AVERAGE (**x**)

$\sum_{i=1}^{n} x_i$ can be found by Authoring n * AVERAGE (**x**)

and similarly for $\bar{y}$.

(iii) use the dot or scalar product to compute $\sum_{i=1}^{n} x_i^2$, etc.,

i.e. $\sum_{i=1}^{n} x_i^2$ can be found by Authoring **x.x**

$\sum_{i=1}^{n} x_i y_i$ can be found by Authoring **x.y** , etc.

The above expressions for $\hat{\beta}_1$ and $\hat{\beta}_0$ can then be Authored as

$$\hat{\beta}_1 := (\mathbf{x}.\mathbf{y} - n * \text{AVERAGE}\ (\mathbf{x}) * \text{AVERAGE}\ (\mathbf{y})) \\ / (\mathbf{x}.\mathbf{x} - n * \text{AVERAGE}\ (\mathbf{x}) * \text{AVERAGE}\ (\mathbf{x}))$$

$$\hat{\beta}_0 := \text{AVERAGE}\ (\mathbf{y}) - \hat{\beta}_1 * \text{AVERAGE}\ (\mathbf{x})$$

For the current data set we obtain from DERIVE

$$\hat{\beta}_0 \simeq 5.197 \ , \quad \hat{\beta}_1 \simeq 0.384$$

Hence the least-squares straight line fit to the lizard data is $y = 5.197 + 0.384\,x$. This line is superimposed on the lizard data plot in Fig. 5.4. Does it look a 'reasonable' fit? Does it match the line you would have fitted 'by eye'?

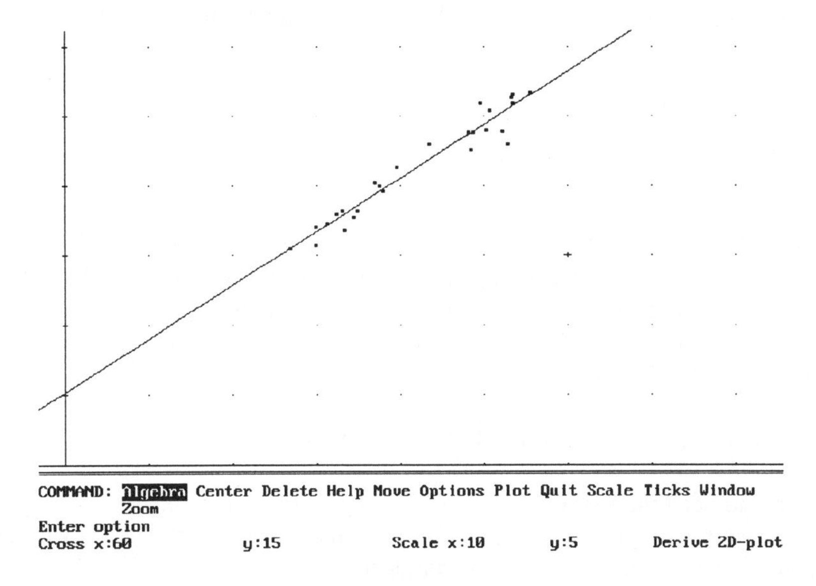

Figure 5.4 Least-squares straight line fitted to lizard data

The distribution of the error term ε is specified if its standard deviation is known. This can be estimated by the formula (see [5] for details):

$$S = \sqrt{\frac{SS_{yy} - \hat{\beta}_1 SS_{xy}}{n-2}} \tag{5.10}$$

where $SS_{yy} = \sum_{i=1}^{n} y_i^2 - \frac{\left(\sum_{i=1}^{n} y_i\right)^2}{n}$ and SS_{xy} is as defined previously.

Using the DERIVE functions described earlier, we obtain $S \simeq 0.924$, for this problem.

Interpretation

Given our straight line model estimator $y = \hat{\beta}_0 + \hat{\beta}_1 x$, it is reasonable to suggest that substituting a specific value for x will result in an estimate for y. But what about the error component? If the straight line model is used to predict the average y value for a given x, then the error component ε does not intrude (since its mean value has been assumed to be zero). It does feature when attempting to predict an individual value of y from a given x, and so to ignore it increases the source of error in the prediction.

A 100(1-α)% prediction interval for an individual y for a given x is given by (see [5])

$$(\hat{\beta}_0 + \hat{\beta}_1 x) \pm t_{\alpha/2,n-2} \,.\, S \,.\, \sqrt{1 + \frac{1}{n} + \frac{(x-\bar{x})^2}{SS_{xx}}} \tag{5.11}$$

where S is given by (5.10) and $t_{\alpha/2,n-2}$ is an appropriate value from the t-distribution with $n-2$ degrees of freedom. For a 95% prediction interval for example, $\alpha = 0.05$ and with $n = 26$ here, the value of t is 2.064. (This value can be found using standard statistical tables or by using the utility function STUDENT(t,v) found in the utility file PROBABIL.MTH . To obtain the t values,

Author STUDENT(t,24) = 0.95 and soLve.

– remember to Load the utility file and work in approXimate mode).

Substitution of this value of t and a particular value of x into (5.11) will give an interval for y within which we are 95% confident the true value will lie.

Validation

A common measure used to assess the strength of a linear relationship between x and y is termed the CORRELATION COEFFICIENT and is defined as

$$r = \frac{SS_{xy}}{\sqrt{SS_{xx}\ SS_{yy}}} \tag{5.12}$$

By definition r lies between -1 and $+1$ and the higher the magnitude of r, the stronger the linear relationship between x and y. For this problem r is computed using DERIVE as 0.97 (approximately) and this value indicates that y tends to increase with increasing x which matches the earlier modelling conclusion.

It must be remembered that this model and its conclusions are based on the data sample given and for values of x within the sample range, the indicators imply a reasonable model. However, it would in general be dangerous to try to extend this model to values of x outside the sample range.

Further Work

1. The straight line fit using least squares can be computed using the FIT function within DERIVE. This function permits an easy comparison of different models as follows. Assuming vector **d** holds the data pairs:

   (a) Author FIT([x, $a + b\,x$], **d**) and Simplify.

   Confirm that your result matches that obtained previously using the intermediate calculations.

   (b) Author FIT([x, a], **d**) and FIT([x, $a + bx + cx^2$], **d**) to produce a model $y = a$ and $y = a + bx + cx^2$ respectively. Plot these models on the scattergram.

   Does the fit appear to improve as more terms are introduced into the model? Do the models show reasonable agreement outside the sample data range?

2. The assumption that the error component is normally distributed leads to the expectation that about 95% of the observations should lie within two standard deviations of the least-squares line, i.e. within $\beta_0 + \beta_1 x \pm 2S$ where S is the estimated standard deviation given by equation (5.10). Use DERIVE to confirm whether this is so in this case.

3. The sample size often affects model predictions significantly. Select a subset of the original data sample with, say, $n = 10$ values with roughly the same range as before. Use the FIT function to compute the regression line. Do your results change significantly? Try varying the sample size and reducing the sample range. What do you conclude?

4. Rather than using the correlation coefficient, r, as a measure of 'goodness of fit', many statistical computer packages produce values of r^2 and '*adjusted* r^2 ' as indicators of 'the worth of x as a predictor for y'. The *adjusted* r^2 value is given by

$$\textit{adjusted } r^2 = 1 - \left(\frac{S}{S_y} \right)^2$$

$$\text{where} \quad S_y^2 = \frac{\sum_{i=1}^{n} (y_i - \bar{y})^2}{n-1}$$

Compute r^2 and *adjusted* r^2 for this problem. The higher their value, the greater the worth of x for predicting y. What do you conclude?

References

[1] Stefani R.T., 'Forever upwards and onwards', *Olympic Review*, **287**, 456–461, IOC (1991).

[2] Stefani R.T., 'Applying least squares to team sports and Olympic winning performances', *Proc. Mathematics and Computers in Sport Conference of the Australian Mathematical Society*, Bond University, Australia, July (1992).

[3] *School Mathematics Project (Advanced Level) Book 4*, pp. 1353–1355, Cambridge University Press.

[4] Alexander D., McClements K. and Simmons J., 'Calculating to win', *New Scientist*, 30–33, December (1988).

[5] Scheaffer R. L. and McClave J.T., *Statistics for Engineers.* PWS Publishers (1982).

6

Modelling With Differential Equations

6.1 INTRODUCTION

This chapter concentrates on the solution of mathematical modelling problems using differential equations. Models are presented which utilize both analytic and numerical methods.

It must be emphasized that a differential equations approach is probably not the only solution to these particular problems but it reflects the solutions developed by the authors. Solutions which lead to the use of alternative mathematical methods can be just as good, or better! Remember, there is no unique answer to a modelling problem.

The early case studies in this chapter record the complete modelling strategy from the original 'brain-storming' session to produce a features list through to discussion of the mathematical solution obtained. Other case studies begin with the authors' pruned features list and develop the solution from there.

6.2 POPULATION DYNAMICS

The Problem

In 1981 the Indian government declared its intention to attempt to limit its population to 1000 million by the end of the century, from its 1981 value of 684 million. It would attempt to achieve this by

(i) reducing the yearly birth rate from 36 per thousand in 1981 to 21 per thousand in the year 2000, by promotion of birth control,

(ii) reducing the yearly death rate from 14 per thousand in 1981 to 9 per thousand in the year 2000, by developing a more effective medical care network.

Develop a mathematical model with which to assess the compatibility of the above aims and use it to assess whether the government's objective is achievable.

Setting up a Model

Obviously, the first thing to notice about the two steps that the Indian government proposes to take is that they are in direct conflict. If health care is improved then more people will live longer so they are likely to produce more children but this is countered by the intention to reduce the birth rate – direct conflict indeed.

There is obviously a myriad of features which will have a bearing on this problem. While no claim is made that it is definitive, the following list represents the authors' feature list:

- initial size of population,
- age distribution of population, as this affects reproduction rates,
- birth rate,
- death rate,
- immigration and emigration rates,
- existence of any governmental support for a family planning policy,
- effect of religious beliefs,
- effect of armed conflicts, natural disasters, ...,
- plus any more YOU can think of.

This list is clearly far too lengthy to form the basis of a viable model and we must thus carry out some fairly severe 'pruning' in order to produce a more manageable pruned feature list:

- initial size of population,
- birth rate,
- death rate.

With this more manageable list, the variables can now be defined and the population growth model can be formulated mathematically.

Formulate a Mathematical Model

We define the variables as

t time (years), with $t = 0$ corresponding to 1981 for convenience

$b(t)$ birth rate (number of births/year/head of population)

$\mathrm{d}(t)$ death rate (number of deaths/year/head of population)

$p(t)$ population size at time t

p_0 initial population size (i.e. the 1981 value)

Points to note with these definitions are as follows:

(i) Units have been carefully specified wherever appropriate.
(ii) A false time origin has been established for numerical convenience.

It might be argued that (ii) is purely 'numerical cosmetics'. That may well be so, but the resultant simplification to the mathematics is well worthwhile.

After all that preliminary discussion, we are now ready to set up our mathematical model.

With reference to the pruned features list, our basic modelling assumption is that the change in the size of the population over a period of time is equal to the number of births in that time minus the number of deaths. Suppose we let the 'period of time' be represented by δt(years), then the modelling assumption can be stated as

Change in size of population from time t years to t + δt years = (*number of births*) − (*number of deaths*)

and expressed mathematically as

$$p(t+\delta t) - p(t) = b(t)p(t)\delta t - d(t)p(t)\delta t \qquad (6.1)$$

i.e.

$$\frac{p(t+\delta t) - p(t)}{\delta t} = \big(b(t) - d(t)\big)p(t) \qquad (6.2)$$

which leads to the differential equation model

$$\frac{\mathrm{d}p}{\mathrm{d}t} = \big(b(t) - d(t)\big)p(t) \qquad (6.3)$$

to be solved subject to the initial condition $p(0) = 684 \times 10^6$.

To progress mathematically from equation (6.2) to (6.3) utilizes the limiting process as $\delta t \to 0$. In fact in this problem the sensible time step to use would be one year so you might argue that δt never even gets close to zero! However $\delta t = 1$ *is* small compared with the nineteen-year time span of interest in the problem and so we can defend the differential equation (6.3) as representing a good mathematical approximation to our modelling assumption.

What can we do, though, if the differential equations approach is not considered valid? For example, for some species of deer (mating once per year), and blue whales (mating once every two years), the time step (one or two years) is large compared with the expected lifetime so that a differential equations approach is inappropriate, as it is based on small changes. In such cases we use the alternative recurrence equation approach

$$P_{n+1} - P_n = \left[b_n - d_n\right] P_n \cdot \Delta t \tag{6.4}$$

where Δt is the time step (one year) and P_k, b_k, d_k represent, respectively, the population, birth rate and death rate in year k ($k = 1, 2, 3, \ldots$) and P_0 denotes the initial population size (684×10^6). As the name implies, equation (6.4) is used recurrently with $n = 0, 1, 2, 3, \ldots$ to generate the values of $P_1, P_2, P_3, \ldots$ in turn (see Chapter 7 for examples of modelling with recurrence equations).

Solution of the Mathematical Model

We now return to the matter of solving the mathematical model developed as equation (6.3). In order to use DERIVE to assist in the solution of differential equations, one must be able to identify the type of differential equation being solved as the methods offered within the utility files are type specific.

Equation (6.3) is a first-order ordinary differential equation of type variables separable. Its solution can be obtained by first loading the utility file ODE1.MTH (containing the collection of methods for Ordinary Differential Equations of order 1) using the following keystrokes

T for Transfer, L for Load, U for Utility

and then, when requested, specifying the file ODE1.MTH. Having identified the form of equation (6.3) as variables separable you can use the Help command and then select U (for Utilities) to search methods offered in ODE1.MTH until you find the appropriate routine (called SEPARABLE) and establish which parameters are required for solution to continue. Alternatively, you may refer to the *DERIVE User Manual*.

The DERIVE steps needed to solve equation (6.3) are shown in Fig. 6.1. In order to progress any further with this solution we need some information about the birth and death rates, b and d. Since only their current and target values are known, we shall need to propose a model of their variation with time. Before doing this, though, let's consider whether or not the Indian government is actually faced with a potential problem.

Suppose the birth and death rates remain at their current values ($b = 0.036$, $d = 0.014$). What then is the population size in the year 2000 ($t = 19$)? This can be quickly answered using the Manage and Substitute commands applied to the solution shown as expression 5 in Fig. 6.1.

With this result highlighted, carry out the keystrokes M for Manage, S for Substitute, and when requested enter the desired values for b, d and t. Finally, type X for approXimate to give the value of the population in the year 2000 as $1.038\,94 \times 10^9$. The target value is thus significantly violated and hence the government can't afford to leave things as they are. Therefore we really do need models for b and d as functions of time. Our experience/general knowledge/expert advice tells us that any changes in b (and d) will be gradual. The simplest model we could adopt for these gradual changes would be a linear model. (Remember the modelling maxim to keep it simple, at least to start with.)

```
1:    "Indian population models.Author and Simplify"

                                              6
2:    SEPARABLE(b - d, p, t, p, 0, 684 10 )

                                          ┌  p ┐
3:    - 6 LN(5) - 2 LN(3) - 8 LN(2) + LN │ ── │ = t (b - d)
                                          └ 19 ┘

4:    "Solve for p,assuming b and d are constant."

                           b t - d t
5:    p = 684000000 ê

                           0.036 19 - 0.014 19
6:    p = 684000000 ê

                          9
7:    p = 1.03894 10
─────────────────────────────────────────────────────────────────
COMMAND: Author Build Calculus Declare Expand Factor Help Jump soLve Manage
         Options Plot Quit Remove Simplify Transfer moVe Window approX
Compute time: 0.1 seconds
Approx(6)                              Free:100%            Derive Algebra
```

Figure 6.1 DERIVE solution of initial model of the Indian poulation problem

Fig. 6.2 shows linear models of the birth and death rates which achieve the target values, starting from their initial values.

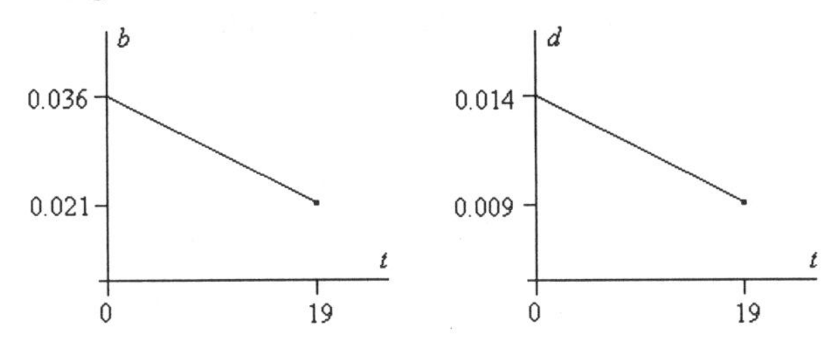

Figure 6.2 Linear models of birth and death rates

Elementary coordinate geometry gives

$$b(t) = 0.036 - \frac{0.015}{19}t$$

and

$$d(t) = 0.014 - \frac{0.005}{19}t$$

Using these expressions the Manage, Substitute strategy applied to expression 2 in Figure 6.1 can be utilized to give the result that, in the year 2000, the population will be $9.447\,85 \times 10^8$ according to this model. The Indian government's objectives have thus been achieved.

If this problem was simply a differential equations 'technique-bashing' exercise, then this is the stage at which you would underline the answer twice, rule off the exercise and carry on to the next one! But wait, modelling isn't like that.

Interpretation and Validation

Is it reasonable to assume that the reductions in the birth and death rates occur at a steady rate (which is the premise on which the linear model was based)? The answer must surely be 'No'. Changes in the birth rate will initially be slow to materialize due to possible delay in the population's acceptance of widespread birth control. Similarly, there will be a lag before there is any evident improvement in the health of the existing population due to improvements in health care. A better mathematical model for b and d would thus be one which included this very slow initial rate of change in both.

We would like a model which gives a very small (zero?) initial rate of change for both b and d while at the same time still giving them their correct initial and final values. Mathematically, we can represent these very small initial rates of change by the extra conditions

$$b'(0) = 0 \quad \text{and} \quad d'(0) = 0$$

zero being an easy approximation for 'very small'! This means that b and d must now each satisfy three conditions, the initial derivative value and the initial and final values. The simplest function which can achieve this is the quadratic function $a + bt + ct^2$, as shown in Fig. 6.3.

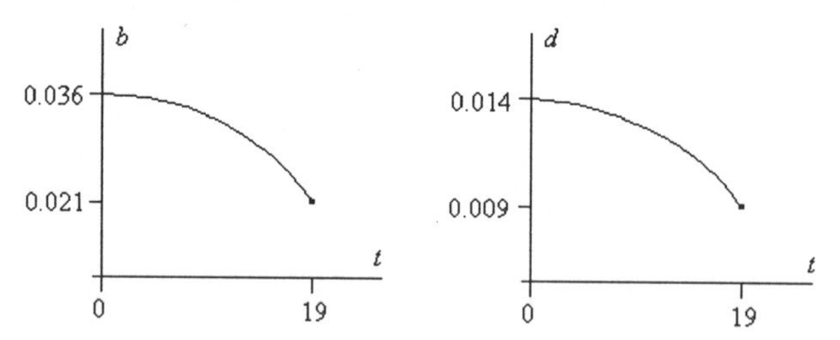

Figure 6.3 More realistic models of birth and death rates

It is left as an exercise for the reader to apply the three conditions to a quadratic model for b (and d) and then to use DERIVE to solve the resulting equations to obtain

$$b(t) = 0.036 - \frac{0.015}{19^2} t^2$$

$$d(t) = 0.014 - \frac{0.005}{19^2}t^2$$

Finally, apply the Manage, Substitute commands to expression 2 of Fig. 6.1 to show that, in this case, the population in the year 2000 is estimated to be $9.751\,82 \times 10^8$.

Not only does the last solution satisfy the government's stated aim but it is based on a model which has a realistic, very slow (actually zero) initial response to improvements in birth and death rates. It thus seems reasonable to conclude that the stated aims of the government are compatible.

Further Work

The model(s) presented here could now be expanded to include more of the original features. For example, if the age distribution of the population were known (say 0–20 years, 20–40 years, ...), then the reproductive patterns of each group could be modelled, thereby leading to a more reliable estimate of the total population in the year 2000. Such data are generally available from a population census.

A Final Thought

You may be wondering how to decide in general whether to opt for a differential equations model or the alternative recurrence equation based approach of equation (6.4). There is unfortunately no hard and fast rule but the differential equations model is the appropriate one if the changes can be considered to be small (as we discussed earlier) even though it may be sensible for the data to be strictly discrete.

6.3 POPULATION DYNAMICS WITH EXPLOITATION

Introduction

This case study is about the modelling of two populations which interact with each other. The problem is presented in terms of foxes and rabbits (the interaction being the fact that foxes feed on rabbits), although it could equally well be interpreted as a model of combat between two opposing forces.

The Problem

Consider the situation in which some foxes and rabbits co-exist in a remote area which has a plentiful supply of food for the rabbits. The foxes depend upon a plentiful supply of rabbits as their food source.

Develop a mathematical model which will show how the two species interact, assuming at the outset that the time step used is small enough to make a differential equations model appropriate.

Setting up a Model

Some thought about the 'mechanics' of the interaction between these two species will help to focus our ideas and avoid our having to produce an exhaustive features list, much of which will be made redundant later. It should be evident to you that in the absence of any foxes the rabbit population will continue to grow, whereas without rabbits the fox population will decline (they will starve to death). From what we know of rabbits, a reasonable model for the growth of their population is an exponential model. Thus if t represents time and $r(t)$ represents the size of the rabbit population at time t, then the rate of change of the rabbit population is given by

$$\frac{dr}{dt} = ar \tag{6.5}$$

where a is a positive constant related to the rate of growth of a rabbit population. You might say that this model should include a term to represent rabbits which die of natural causes. We shan't include this in view of the basic modelling principle: 'keep it simple'! Anyway, once the foxes start preying on the rabbits they may not live long enough to die of old age!

As mentioned above, the fox population would decline if there were no rabbits and we could assume an exponential model for this, too (other models are possible of course but this one has the virtue of simplicity). Hence if $f(t)$ denotes the size of the fox population at time t then

$$\frac{df}{dt} = -bf \tag{6.6}$$

where b is a positive constant related to the rate of decline of the fox population (the minus sign ensures a decline).

Formulate a Mathematical Model

Obviously, we now need to model the interaction, because common sense tells us that if there is an abundance of rabbits then the foxes will feed well and multiply. If, however, a large fox population eats too many rabbits, then the rabbit population will decline causing a reduction in the number of foxes later, which will give the rabbits a chance to re-establish themselves, and so on. We would not be surprised, therefore, if the solution to our problem exhibited a cyclic, or periodic behaviour. Before we can attempt to produce this solution we need to model the interaction between the foxes and the rabbits.

How can we represent the above interaction mathematically? We need a function of r and f with the following properties:

(i) For a fixed value of f the number of rabbits killed increases as r increases (easier to kill if they are plentiful).
(ii) For a fixed value of r the number of rabbits killed increases as f increases.

The simplest expression with these properties is rf and is incorporated as follows into the model represented by equations (6.5) and (6.6) to give

$$\frac{\mathrm{d}r}{\mathrm{d}t} = ar - crf \tag{6.7}$$

$$\frac{\mathrm{d}f}{\mathrm{d}t} = erf - bf \tag{6.8}$$

The term crf (c is another positive constant) occurs as a negative contribution since it represents a further reduction of rabbits, while in the fox equation (6.8) the interaction term erf (e is another positive constant) occurs as a positive contribution as it represents their increase, depending upon the number of rabbits killed.

Solution of the Mathematical Model

We are now in a position to begin the solution to our mathematical model. The following parameter values have been found to be representative [1]:

$a = 2$

$b = 0.45$

$c = 0.01$: represents the proportion of the rabbit population killed by one fox per unit time

$e = 0.0005$: represents the proportion of extra births of foxes per unit time due to the food value of one rabbit

and we shall assume that initially there are 800 rabbits and 100 foxes, i.e.

$$\text{at } t = 0, \quad r = 800 \quad \text{and} \quad f = 100 \tag{6.9}$$

Recall that in this problem we are more interested in how the numbers of foxes and rabbits depend on each other, rather than how each varies with time. Accordingly, we shall eliminate t between equations (6.7) and (6.8) to obtain

$$\frac{dr}{df} = \frac{2r - 0.01rf}{0.0005rf - 0.45r} \tag{6.10}$$

to be solved subject to the initial conditions (6.9). The presence of the *rf* term renders equation (6.10) non-linear, and we must use a numerical method to obtain the solution. (Readers should convince themselves that the utility file ODE1.MTH does not contain a function likely to solve equation (6.10).) There are, of course, many numerical methods available for the solution of ordinary differential equations and one of the most commonly used is the Runge–Kutta method of order four, which is available within the DERIVE utility file ODE_APPR.MTH, and is used here to soLve equations (6.7) and (6.8), with the suggested parameter values and the initial conditions being given in equation (6.9).

The following sequence of keystrokes will load this particular file

T for Transfer, L for Load, U for Utility

and then, when requested, specify the file name as ODE_APPR.MTH.

To implement the Runge–Kutta method, first Author expression 2 in Fig. 6.4, which shows the various steps of the Runge–Kutta solution (we used sixty steps of size 0.25).

```
1:   "Rabbits and foxes numerical solution"

2:   RK([2 r - 0.01 r f, 0.0005 r f - 0.45 f], [t, r, f], [0, 800, 100], 0.25, 6

3:   "approXimate to give the (t,r,f) solution"

4:   [[0, 800, 100], [0.25, 1027.67, 100.114], [0.5, 1314.64, 103.501], [0.75, 1

5:   "Extract the (r,f) data from the solution"

6:   EXTRACT_2_COLUMNS([[0, 800, 100], [0.25, 1027.67, 100.114], [0.5, 1314.64,

7:   "and approXimate to obtain"

8:   [[800, 100], [1027.67, 100.114], [1314.64, 103.501], [1658.84, 111.319], [2

COMMAND: Author Build Calculus Declare Expand Factor Help Jump soLve Manage
         Options Plot Quit Remove Simplify Transfer moVe Window approX
Compute time: 0.8 seconds
Approx(6)                                Free:85%                  Derive Algebra
```

Figure 6.4 DERIVE's Runge–Kutta solution of the rabbits and foxes problem

In order to produce a graph of *f* plotted against *r* we must first extract the (*r*,*f*) data from our solution. This is achieved with expression 6 of Fig. 6.4, namely,

EXTRACT_2_COLUMNS(#4, 2, 3)

since the (*r*,*f*) data are held in columns 2 and 3 of the results matrix represented by expression 4. Finally, the approXimate command has been used to give the results in expression 8.

These results can then be presented graphically using the Plot command. With the (r, f) results highlighted, type P (for Plot) and the screen will be refreshed with a set of axes. Before plotting the results, drive the cursor to somewhere central in the first quadrant and type C (for Centre). The screen will now contain a set of axes presenting mainly the first quadrant. (Why did we do this? Because r and f are each strictly non-negative.)

Next choose some appropriate scales (we'll save you time here by suggesting 'x scale' of 750 and 'y scale' of 150, although generally you would have to experiment with these values for yourself until you were satisfied with the presentation). Finally, type P (for Plot) when the (r, f) results highlighted earlier will be presented, see Fig. 6.5, in which rabbits are plotted horizontally and foxes vertically. The graph you have just obtained is called a phase plane diagram.

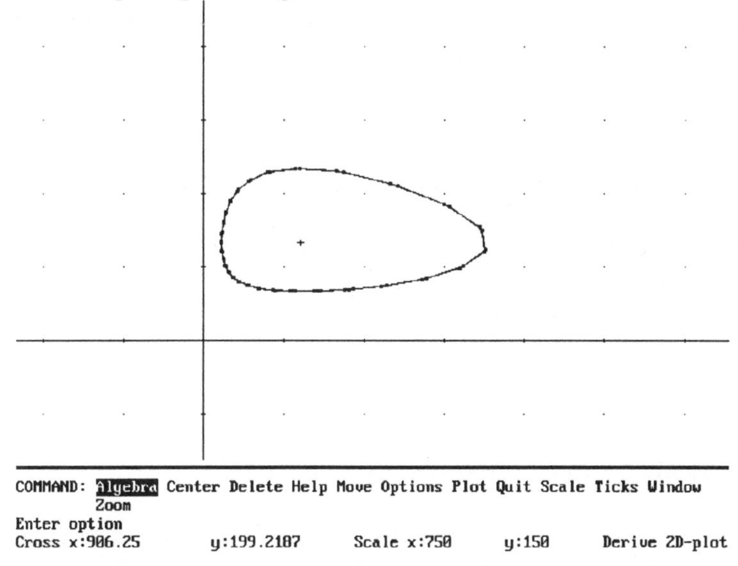

Figure 6.5 Predator-prey solution number – number of foxes plotted vertically against number of rabbits

Interpretation and Validation

Note that there is a cyclic behaviour to the population sizes as we hypothesized at the outset – a case of the model being validated by our knowledge/experience.

What else can we find out about these animals? Can they ever live happily together? This would mean their numbers remaining constant, which we can express mathematically as

$$\frac{dr}{dt} = 0 \quad \text{and} \quad \frac{df}{dt} = 0$$

Solution of these two equations then gives (see equations (6.7) and (6.8))

$$r = b/e = 900$$

$$f = a/c = 200$$

This is why the cross-hair is set as closely as possible to this position in Fig. 6.5. Notice that horizontal and vertical lines through this equilibrium position divide the first quadrant into four regions, within which the trajectory (the name of the curve plotted in the phase plane) has four distinct phases which the reader is encouraged to identify as follows:

(i) Many rabbits, and the foxes increase so reducing rabbit population.
(ii) Foxes and rabbits decline.
(iii) Few foxes now so rabbits re-established.
(iv) Sufficient rabbits for foxes to increase again.

These four stages agree with our expectations and so provide a further measure of validation.

We mentioned earlier that we might expect the rabbit (and fox) populations to be periodic with time. Well, let's see! The Runge–Kutta results are presented as triplets $[t,r,f]$, from which the $[t,r]$ and $[t,f]$ data can be extracted by Authoring and Simplifying the appropriate expressions

EXTRACT_2_COLUMNS(#4, 1 ,2) , for $[t,r]$

EXTRACT_2_COLUMNS(#4, 1 ,3) , for $[t,f]$

Fig. 6.6 shows these two sets of results (note the use of Centre to focus on the first quadrant and note, too, the scales we used, 'x scale' 2 and 'y scale' 750) and the anticipated periodic behaviour is at once apparent, giving yet more validation. Notice, too, that neither population ever dies out at any stage and that there is a much greater variation in the rabbit population.

Further Work

Some suggestions for further work:

- How sensitive are the solutions to variations in the values of the parameters a, b, c and e?
- Incorporate a crowding term into equations (6.7) and (6.8) to make the model more realistic. (Crowding can be modelled by r^2 and f^2, can you explain why?)

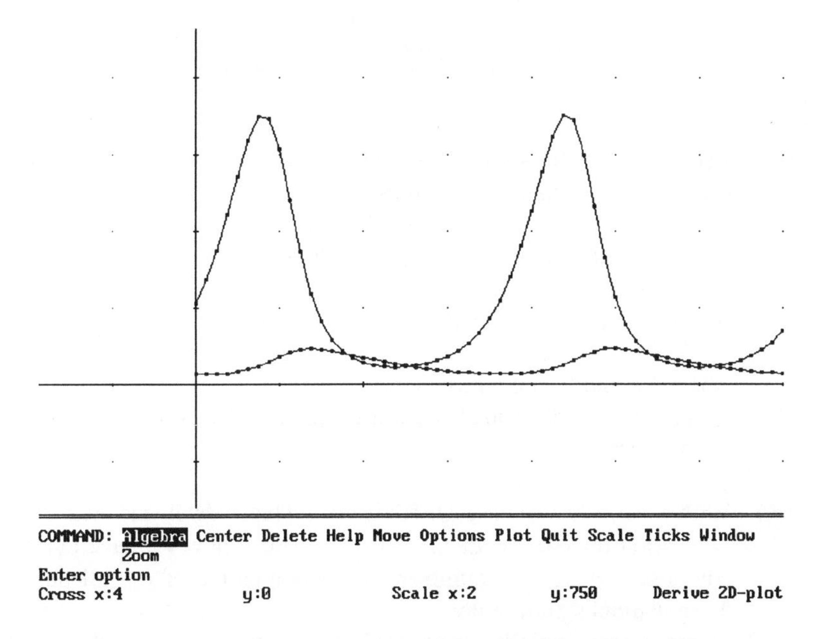

Figure 6.6 Time variation of fox and rabbit population

A related exercise, see reference [2], which you may wish to investigate is as follows. The population sizes x and y of two species who compete for their food supply can be modelled by the differential equations

$$\frac{\mathrm{d}x}{\mathrm{d}t} = ax - by \tag{6.11}$$

$$\frac{\mathrm{d}y}{\mathrm{d}t} = cy - dx \tag{6.12}$$

where a, b, c and d are each positive constants.

(i) From a modelling point of view can you explain the significance of each of the terms on the right-hand sides of the equations?

The next sections illustrate an alternative approach to the solution of pairs of differential equations such as (6.11) and (6.12).

(ii) By differentiating equation (6.11) once with respect to time and then using (6.12) to replace the $\mathrm{d}y/\mathrm{d}t$ term, show that x satisfies the second-order differential equation

$$\frac{d^2x}{dt^2} - (a+c)\frac{dx}{dt} + (ac-bd)x = 0$$

(iii) Use DERIVE to show that the general solution of this differential equation for the parameter values

$$a = c = 3 \quad ; \quad b = d = 2$$

is $$x = C_1 e^{5t} + C_2 e^t$$

where C_1, and C_2 are arbitrary constants.

You will need to load the utility file ODE2.MTH, and then use the function DSOLVE2.

(iv) With x now known, use DERIVE to evaluate dx/dt and hence find an expression for y from equation (6.11). Notice how we have managed to completely avoid integration in the determination of y – all part of the 'keep it simple' philosophy.

(v) Suppose that initially (at $t = 0$) $x = 100$ and $y = 200$. Use this information to obtain the values of C_1 and C_2.

(vi) Use the soLve facility to determine the time at which one of the species (which one?) becomes extinct.

6.4 DRUG TRANSPORT

Introduction

Mathematical modelling is widely used in the pharmaceutical industry to develop models from which researchers can either predict the likely performance of a proposed new drug or possibly suggest a new design. Obviously, any new drugs still have to go through a rigorous series of laboratory tests and clinical trials but in the early stages of the development of a new drug, modelling may be able to save a drug company a great deal of time and money.

This case study is developed from work carried out by one of the authors to develop a model of the drug transport process, drug transport being the name used to describe the movement of a drug through the body from its point of introduction to the body to its desired site of action/excretion, see reference [3].

The Problem

The transport is assumed to take place through many compartments, alternately aqueous and lipid, each separated from its neighbour by a thin porous membrane. For simplicity a linear path is assumed, of the form shown in Fig. 6.7 (a). Here C_i denotes the concentration of the drug in the ith compartment at time t. We shall assume that a unit dose is administered at time zero in compartment 1, i.e.

$$C_1 = 1 \quad \text{and all other} \quad C_i = 0 \quad \text{at} \quad t = 0$$

The constants k and ℓ are known as the forward (aqueous to lipid) and reverse (lipid to aqueous) partitioning rates and describe the rate of loss/gain of concentration in each compartment. (The rate of change of C_1 is reduced by an amount kC_1, but augmented by ℓC_2 from the second compartment.) The final coefficient m only acts from C_{n-1} to C_n i.e. the final membrane is irreversible.

At this stage you may be thinking 'But this is all chemistry and I'm not interested in that.' Perhaps so, but often mathematical modellers have to formulate problems presented by disciplines alien to them. In such cases it is invaluable to be able to have discussions with people who are knowledgeable in that particular field. Such discussions will often help the modellers to identify the key variables and begin to focus on possible and feasible models. Without the background information and/or dialogue the modellers could well end up with models which are very attractive mathematically but absolutely useless for the required purpose!

Setting up a Model

After their discussions with the experts, the modellers realized that the problem was essentially an input–output situation for each compartment. The assumption was therefore made that the change in concentration in each compartment was proportional to the net gain/loss across the bounding membranes; mathematically speaking, the partitioning rates are the constants of proportionality. Although the original model consisted of many compartments, the three-compartment model investigated here (see Figure 6.7(b)) is merely a simpler case which still illustrates all the modelling aspects of the original work. Further, this initial model assumes that the membranes only permit flow from left to right, i.e. that they are irreversible.

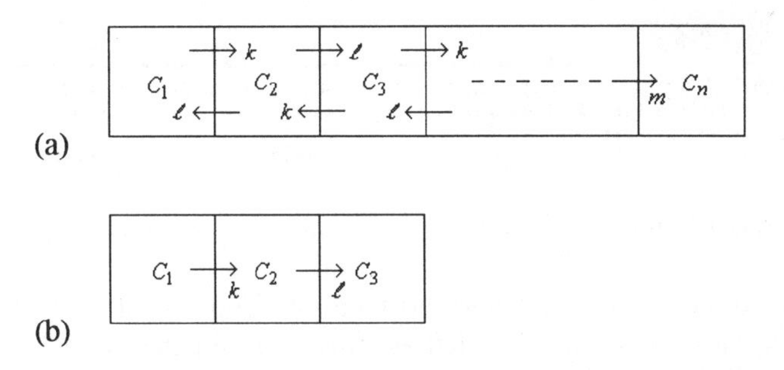

Figure 6.7 (a) Multistage linear model of the drug transport process
(b) a simplified three compartment model

Formulate a Mathematical Model

Applying the input–output principle to each compartment, the equations governing the concentrations for this three-compartment model are found to be

$$\frac{dC_1}{dt} = -kC_1 \tag{6.13}$$

$$\frac{dC_2}{dt} = kC_1 - \ell C_2 \tag{6.14}$$

$$\frac{dC_3}{dt} = \ell C_2 \tag{6.15}$$

These three differential equations each possess an analytic solution. It is worthy of note for tutors that use of a package such as DERIVE does not remove the need for the student to be able to recognize the different types of differential equation as the utility file ODE1.MTH lists its methods of solution according to the classification of the first-order differential equation.

Solution of the Mathematical Model

Equation (6.13) is of the type 'variable separable'. Its solution is detailed below in Fig. 6.8.

```
1:   "Solution of the differential equation for c1.Author"

2:   SEPARABLE(-k, c1, t, c1, 0, 1)

3:   "and Simplify to give"

4:   LN(c1) = - k t

5:   "SoLve for c1 to give"

           - k t
6:   c1 = ê
```

```
COMMAND: Author Build Calculus Declare Expand Factor Help Jump soLve Manage
         Options Plot Quit Remove Simplify Transfer moVe Window approX
Compute time: 0.1 seconds
Solve(4)                              Free:91%              Derive Algebra
```

Figure 6.8 DERIVE solution for the concentration in the first compartment

With C_1 now known, it can be substituted into equation (6.14), which can then be identified as 'first-order linear'. It is left as an exercise for the reader to show that the solution for C_2 is given by

$$C_2 = \frac{ke^{-kt}}{\ell - k} + \frac{ke^{-\ell t}}{k - \ell}$$

Finally, C_3 can be obtained by integrating the now known C_2 and using the initial condition that $C_3 = 0$ at $t = 0$ to give

$$C_3 = \ell\left[\frac{e^{-kt}}{k-\ell} + \frac{ke^{-\ell t}}{\ell(\ell - k)} + \frac{1}{\ell}\right]$$

Interpretation

The solutions for C_1, C_2 and C_3 can now be plotted for different values of k and ℓ. For illustrative purposes output for the following three cases is presented in Figs. 6.9–6.11 respectively:

(i) $k = 2$, $\ell = 0.5$

(ii) $k = 0.5$, $\ell = 2$

(iii) $k = 1$, $\ell = 1$

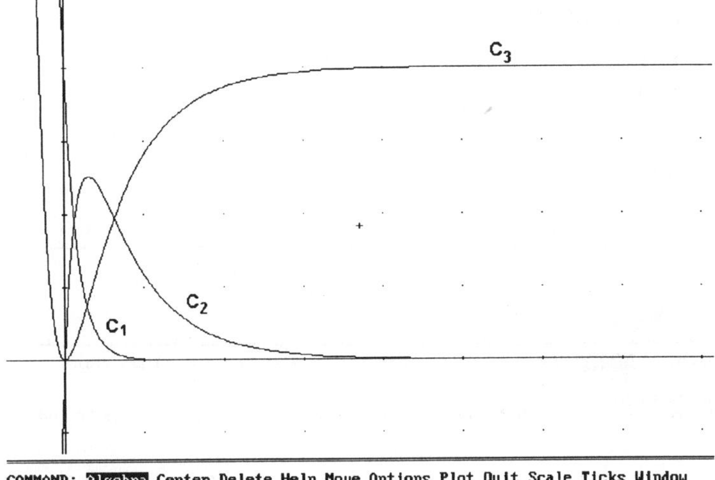

Figure 6.9 Time variation of C_1, C_2 and C_3 for the case $k = 2$, $\ell = 0.5$

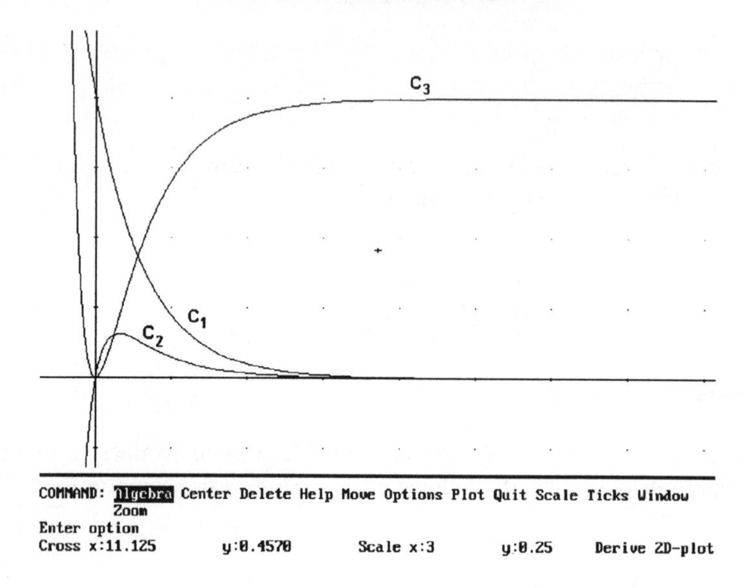

Figure 6.10 Time variation of C_1, C_2 and C_3 for the case $k = 0.5$, $\ell = 2$

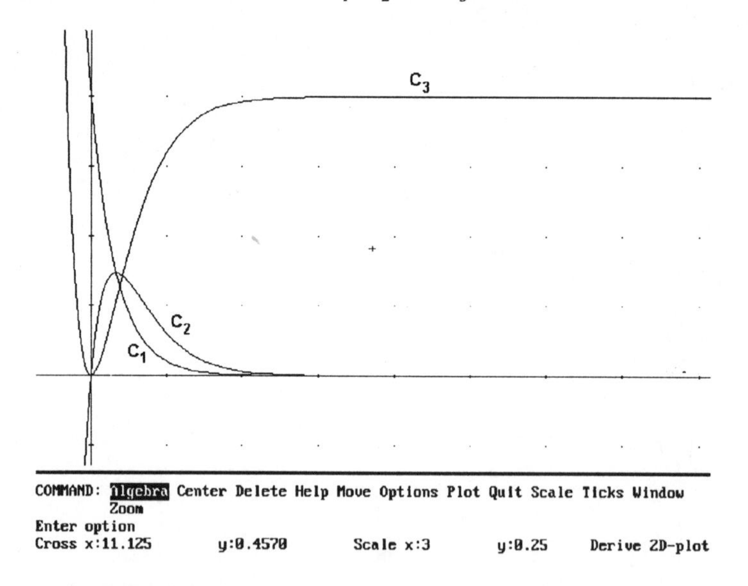

Figure 6.11 Time variation of C_1, C_2 and C_3 for the case $k = 1$, $\ell = 1$

Note that the case where $k = \ell = 1$ is a special case (Can you see why? Hint: look at the expressions for C_2 and C_3 and notice the factor $(k - \ell)$ in the denominator.) Solution of this special case is left as an exercise for the reader.

The graphical output shows that although the basic shapes are preserved, the gradients and maxima are dependent on the values of k and ℓ. Thus, by driving the cross-hair onto the appropriate point on the solution curve the modeller can answer questions such as 'How long does it take for C_2 to reach a maximum for a specific set of values of k and ℓ?', 'How long does it take for C_3 to reach 90% of its maximum?', and so on.

A Revised Model

At this stage, it is worth noting that in reality the intercompartmental membranes actually permit flow in both directions and so a more reasonable model would be that illustrated in Fig. 6.12. The final membrane is assumed to be irreversible – you don't want the drug to migrate from its site of action once it has got there!

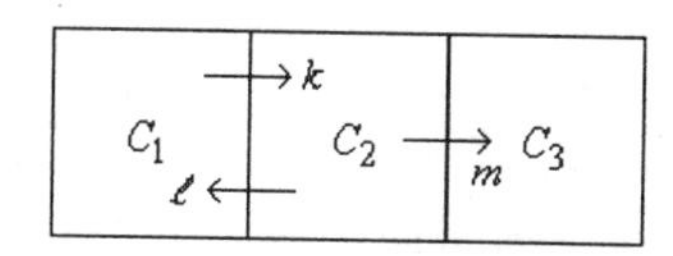

Figure 6.12 Three compartment model incorporating bi-directional flow

The input–output model then gives

$$\frac{dC_1}{dt} = -kC_1 + \ell C_2 \tag{6.16}$$

$$\frac{dC_2}{dt} = kC_1 - (\ell + m)C_2 \tag{6.17}$$

$$\frac{dC_3}{dt} = mC_2 \tag{6.18}$$

This set of differential equations is known as a coupled set, as C_1 can't be found unless C_2 is known (see equation (6.16)) but C_2 can't be found unless C_1 is known (see equation (6.17)). Such systems are most readily solved using a numerical method. By typing the command Help followed by Utility, the reader can then browse through the contents of the various utility files available within DERIVE. Under the heading of first-order ordinary differential equations there is a utility file named ODE_APPR.MTH which deals with the approximate solution of such equations. Within this the user is directed to the fourth-order Runge–Kutta method for the numerical solution of systems of first-order ordinary differential equations. To effect the solution of equations (6.16)–(6.18), first substitute example values of k, ℓ, and m (say 2, 0.5 and 1 respectively) and then load the file ODE_APPR.MTH (Transfer, Load, Utility, followed by the file name). Next, Declare t, C_1, C_2 and C_3 as non-negative variables.

The Runge–Kutta solution is shown in Fig. 6.13. After some experimentation we found it necessary to let the time t run from 0 to 20 in steps of 0.5 in order for C_3 to reach a maximum and stabilize.

In order to produce a graphical output of (t, C_1), (t, C_2) or (t, C_3) we need to extract the appropriate data from the complete solution matrix. This is achieved by Authoring and approXimating the expression

EXTRACT_2_COLUMNS(#4, 1, 2)

to extract columns 1 and 2 from expression 4 of Fig. 6.13 to give the set of (t, C_1) data, see expressions 6 and 7 of Fig. 6.13. (The modifications for the (t, C_2) and (t, C_3) data are self-evident.)

```
1:   "Runge-Kutta solution of the drug transport problem."

2:   RK([- 2 c1 + 0.5 c2, 2 c1 - 1.5 c2, c2], [t, c1, c2, c3], [0, 1, 0, 0], 0.5

3:   "Now approXimate to give the solution"

4:   [[0, 1, 0, 0], [0.5, 0.440755, 0.401692, 0.157552], [1, 0.234604, 0.394435,

5:   "Extract the (t,c1) results using"

6:   EXTRACT_2_COLUMNS([[0, 1, 0, 0], [0.5, 0.440755, 0.401692, 0.157552], [1, 0

7:   [[0, 1], [0.5, 0.440755], [1, 0.234604], [1.5, 0.143013], [2, 0.0939340], [

COMMAND: Author Build Calculus Declare Expand Factor Help Jump soLve Manage
         Options Plot Quit Remove Simplify Transfer moVe Window approX
Compute time: 0.3 seconds
Approx(6)                                Free:95%                Derive Algebra
```

Figure 6.13 DERIVE's Runge–Kutta solution for the bi-directional flow model

Once the three sets of data have been extracted they can then be presented graphically using the Plot command, see Fig. 6.14.

(Suitable scales for the parameter values used were found to be 'x scale' 2.5 and 'y scale' 0.25.)

Interpretation and Validation

It is interesting to note that the general shape of the solution curves is similar to the non-reversible membranes model used initially. We are therefore led to ask the question 'Was all this extra work necessary/worthwhile?' To answer this, Plot results for C_2 for the two models (for the case $k = 2$, $\ell = 0.5$, $m = 1$) and notice that the two maxima are significantly different both in their value and the time taken for them to be attained.

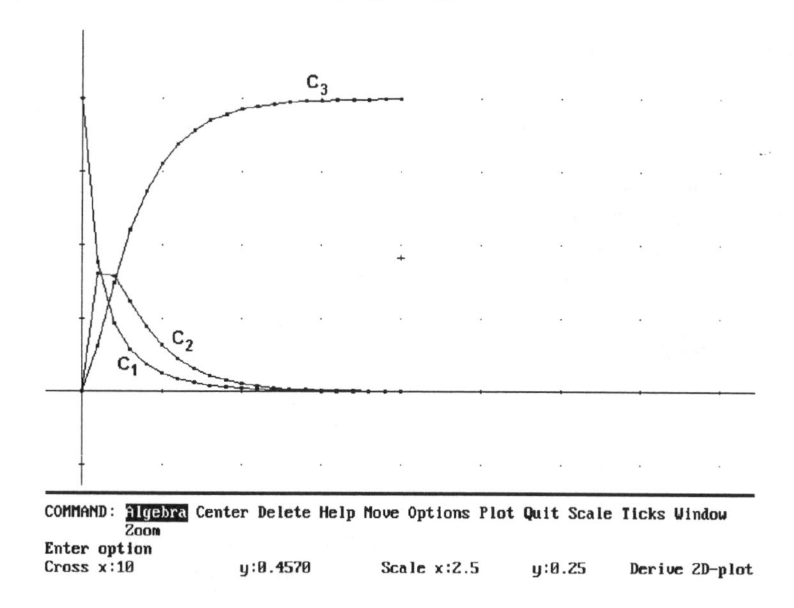

Figure 6.14 Time variation of C_1, C_2 and C_3 for the bi-directional flow model and the case $k = 1$, $\ell = 1$, $m = 1$

There are therefore strong grounds for adopting the results from the revised, more realistic model (reversibility at the membranes).

At this point the pharmacists might take over and carry out some laboratory experiments to measure the various concentrations and hopefully thereby validate the model, as happened with the authors' original work.

In conclusion, the differential equations (6.16)–(6.18) possess an interesting mathematical property. If a Runge–Kutta solution is attempted for values of k and ℓ which are orders of magnitude different (say 1000 and 0.001), as is the case for some drugs, then nonsense results are obtained (some of the concentration values are negative!). This is because the various component mathematical solutions are changing at vastly different rates and the Runge–Kutta method cannot approximate this behaviour adequately. The numerical solution is said to exhibit instability and the differential equations are said to constitute a stiff system. Specialist numerical methods exist for the solution of such systems.

Although this case study has been simplified from the original piece of research work in which it arose, it highlights some valuable modelling lessons:

- the need for ongoing dialogue with an expert,
- the need to validate models (perhaps via experimental work as in this study),
- the value of asking yourself at any early stage whether you've seen anything similar before (this was another input–output model),
- the importance of identifying any special cases ($k = \ell$ in this study).

Further Work

Experiment with other values of k and ℓ (it is recommended that they satisfy $k \cdot \ell = 1$ in order to avoid idiosyncratic results when using DERIVE).

What happens if you change the value of m? How are the solutions affected if a sustained dose is applied to compartment 1 (i.e. $C_1 = 1$ for all time, so $dC_1/dt = 0$)? What happens if one of the compartments has a different volume from the others?

6.5 PROJECTILES WITH SPIN

Introduction

It is well known that in the absence of air resistance and spin effects a projectile follows a parabolic trajectory. The formula for the range attained by the projectile (be it a military shell or a sports implement) is similarly well documented [4]. The range depends on three parameters: the release velocity, the release angle and the release height.

Obviously, varying any (or all) of these will affect the range, but which has the greatest effect? A simple investigation of the effect on the range of a $\pm$ 1% variation in each (taken one at a time, why?) shows that the range is most affected by changes in the release velocity. Mathematically, we might have foreseen this as the range is dependent on the square of the velocity and hence any changes in velocity, although small, are made more significant.

Incidentally, the result explains why shot-put coaches exhort their athletes to concentrate on achieving as great a release velocity as possible. It also explains why shot-putters are very heavy people so that a great deal of momentum is available to be transferred to the shot at release, thereby giving it a large release velocity. This is itself another possible model to investigate.

The Problem

The nature of the release of a projectile (firing, kicking, striking, ...) usually results in some spin being imparted to it. In some cases this is both deliberate and essential: the spin about the longitudinal axis of a bullet, shell or javelin gives each its stability during flight. But what about the spin applied to a sports ball (tennis, cricket, golf, ...) by the action of the striking implement? In this exercise we develop a model for predicting the effect of such spin.

Fig. 6.15(a) shows a vertical section through a spherical sports ball which we shall assume is travelling in the vertical plane of the page, say the xy plane. With the z axis perpendicular to the page, two types of spin about the z axis are considered. These are topspin (if the sense of rotation as viewed by the reader is clockwise) and backspin (anticlockwise as viewed by the reader).

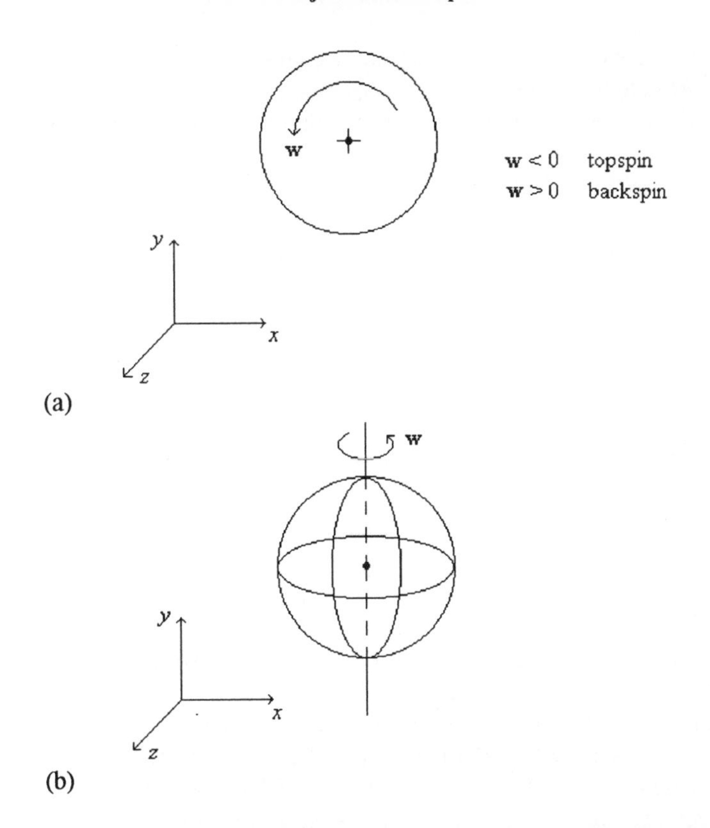

Figure 6.15 (a) Topspin/backspin of a ball moving in the xy plane
(b) The spin required to move the ball out of the xy plane

Alternatively the spin might be about the y axis, which would cause the ball to swerve to the left or right of its initial plane of projection, see Fig. 6.15(b). Generally of course, the ball could be caused to spin about any axis. In order to keep the model of spin reasonably simple we shall restrict our attention to the development of a model covering the topspin/backspin situations shown in Fig. 6.15(a).

Setting up a Model

Referring to Fig. 6.16 the ball is released from position P with an initial translational velocity, in the xy plane, of

$$\mathbf{V}_0 = u_0\mathbf{i} + v_0\mathbf{j}$$

at an angle α to the horizontal and with a spin $\mathbf{w}$ given by

$$\mathbf{w} = w\,\mathbf{k}$$

where w can be positive or negative to accommodate the cases of backspin and topspin respectively. Although for simplicity it is assumed that $|\mathbf{w}|$ is constant, in reality its value would decrease as time progressed due to the effects of friction between the ball and the air.

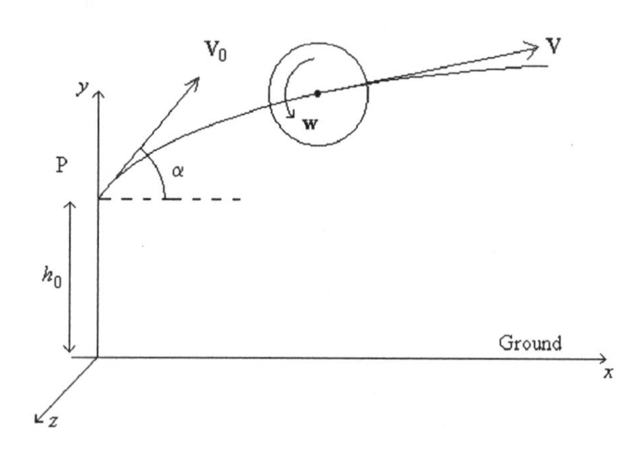

Figure 6.16 Showing the release conditions of a ball moving with spin w in the xy plane

It seems reasonable to suggest that the magnitude of the spin force experienced by the ball will be dependent on the magnitude of both its translational and rotational velocities. Experience also tells us that in the case of topspin, the spin force ($\mathbf{F}_s$) is directed towards the ground, whereas for backspin it is directed upwards. The above assumption and observation are both expressed within the vector model

$$\mathbf{F}_s = \lambda(w \times \mathbf{V})$$

where λ is a constant. (This model was originally proposed by Mooney and is reported by Hughes [5] .)

Formulate a Mathematical Model

The complete mathematical model of the spinning ball can then be expressed using Newton's second law of motion as

$$m\mathbf{a} = -mg\,\mathbf{j} + \lambda(\mathbf{w} \times \mathbf{V})$$

where m is the mass of the ball, $\mathbf{a}$ is its acceleration vector and $g = 9.81$ m s^{-2} is the acceleration due to gravity.

For the topspin/backspin being considered here we have

$$\mathbf{w} = w\mathbf{k}$$

and

$$\mathbf{V} = u\,\mathbf{i} + v\,\mathbf{j}$$

so that

$$\mathbf{w} \times \mathbf{V} = -\lambda wv\,\mathbf{i} + \lambda wu\,\mathbf{j}$$

(Both topspin and backspin are addressed by changing the sign of w.)

In component form, the two equations of motion are thus

$$m\frac{\mathrm{d}u}{\mathrm{d}t} = -\lambda wv$$

and

$$m\frac{\mathrm{d}v}{\mathrm{d}t} = \lambda wu - 9.81$$

When dealing with projectile problems we are usually interested in the trajectory so rather than solve the above differential equations for u and v we would prefer to have solutions for the position co-ordinates x and y. Thus we are faced with the solution of the following set of four first-order differential equations:

$$\frac{\mathrm{d}x}{\mathrm{d}t} = u \tag{6.19}$$

$$\frac{\mathrm{d}y}{\mathrm{d}t} = v \tag{6.20}$$

$$\frac{du}{dt} = -\left(\frac{\lambda w}{m}\right)v \tag{6.21}$$

$$\frac{\mathrm{d}v}{\mathrm{d}t} = \left(\frac{\lambda w}{m}\right)u - 9.81 \tag{6.22}$$

which are to be solved subject to the initial conditions that

$$\text{at } t = 0\,,\ x = 0\,;\ y = h_0\,;\ u = V_0 \cos\alpha\,;\ v = V_0 \sin\alpha$$

The question as to what value to use for λ has been addressed by Mooney (see reference [5]), who reports that for sports projectiles best solutions are obtained by choosing $|\lambda w/m| = 0.2$.

Solution of the Mathematical Model

The coupled nature of the differential equations (6.19)–(6.22) precludes the determination of an analytic solution and so resort has to be taken to numerical methods. Once again, an inspection of DERIVE's utility files leads us to the function RK, which applies the fourth-order Runge–Kutta method to a system of first-order differential equations, for a user supplied step-size and number of steps.

To use Runge–Kutta with DERIVE you must first Transfer and then Load the utility file ODE_APPR.MTH. The Runge–Kutta solution to this problem, is shown in Fig. 6.17, in which the parameter c denotes the quantity $\lambda w/m$ (so that $c = \pm 0.2$ gives the results for the two cases of backspin and topspin) and for which the initial conditions were representative of a baseline drive in tennis

$$h_0 = 1\,\text{m}\,;\; V_0 = 21\,\text{m}\,\text{s}^{-1}\,;\; \alpha = 14°$$

so that initially

$$u = 20.38\,\text{m}\,\text{s}^{-1} \quad \text{and} \quad v = 5.08\,\text{m}\,\text{s}^{-1}$$

```
1:   "RK solution of the spin problem,note c=-0.2,+0.2,0.0"

2:   RK([u, v, - c v, c u - 9.81], [t, x, y, u, v], [0, 0, 1, 20.38, 5], 0.2, 11

3:   "Manage,Substitute c=+0.2 and approXimate..."

4:   RK([u, v, - 0.2 v, 0.2 u - 9.81], [t, x, y, u, v], [0, 0, 1, 20.38, 5], 0.2

5:   "This expression will contain the soln [t,x,y,u,v]"

6:   "Extract the (x,y) data from this solution using"

7:   "the command EXTRACT_2_COLUMNS(#5,2,3) and approXimate..."

8:   "This expression will contain the (x,y) trajectory data."

COMMAND: Author Build Calculus Declare Expand Factor Help Jump soLve Manage
         Options Plot Quit Remove Simplify Transfer moVe Window approX
Enter option
User                    H:TSM.MTH             Free:90%           Derive Algebra
```

Figure 6.17 Numerical solution for the trajectory of a spinning ball

The selected step-size $h = 0.2$ and the number of steps $N = 11$ represent the optimum outcome of several solution attempts. We had to estimate the likely flight time, say 2 s, and then divide it into a reasonable number of steps (say ten). Our first solution finished in mid-air (!) so we had to revise the number of steps! We also ran the problem several times with different step-sizes in order to confirm our solution. You will see from Fig. 6.18 that the trajectories now finish below ground, but at least this means that the projectile has landed!

More sophisticated numerical software packages permit you to easily monitor the value of y and stop the program once y becomes less than or equal to zero. To be fair, DERIVE does permit some programming using the IF construct, but the 'experimental' approach outlined here is the simplest one to use.

The solutions for the three cases $c = 0.2$, -0.2 and 0 (backspin, topspin and no spin respectively) are presented in the form of matrices, each row of which gives a set of values of t, x, y, u and v.

In order to produce graphical output of the trajectories we must plot y versus x for each case. These data can be extracted from the solution matrices by using the EXTRACT_2_COLUMNS command (this is also in the utility file ODE_APPR.MTH, which fortunately has already been loaded). Author the expression

EXTRACT_2_COLUMNS(#5, 2, 3)

and Simplify it to produce from expression 5 (the solution for the case where $c = 0.2$) the contents of the second and third columns (i.e. the (x,y) values). This can then be plotted in the usual way. The trajectories for each of the spin cases (top-, back- and none) are shown in Fig. 6.18, where it is at once evident that the spin has had a significant effect on the trajectory of the ball. (The trajectories have been obtained as continuous curves by using the commands Option, State, Continuous.)

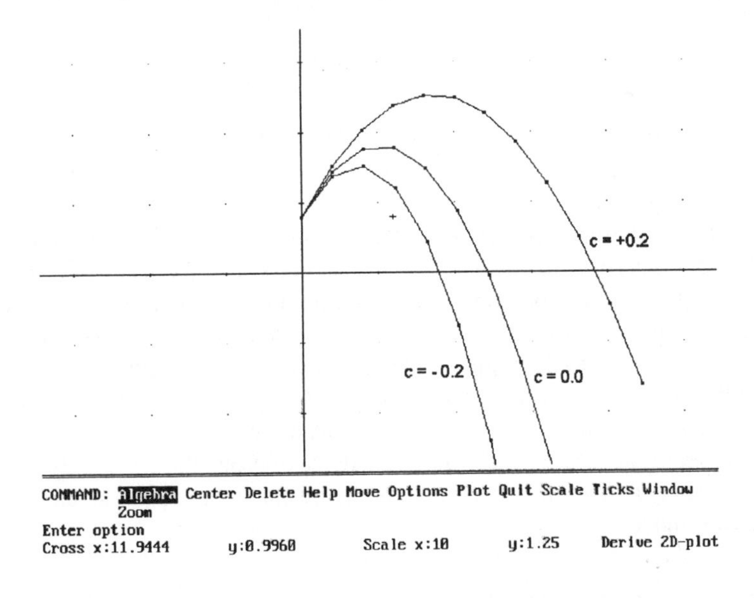

Figure 6.18 Showing the effect of spin on the trajectory of a sports ball

Interpretation

As mentioned previously, the values used as the initial conditions are representative of a baseline drive in tennis. Assume that the net has a height of 1 m and is situated some 12 m away from the release point of the ball. The point representing the top of the net is shown by the cross-hair on Fig. 6.18. Notice how the effect of spin significantly changes the clearance over the net, thereby affecting an opponent's ability to return the shot. The interested reader might research values of h, V_0 and α for other selected shots/other sports to further investigate the effects of spin.

6.6 PROJECTILES AND AIR RESISTANCE

Introduction

Of course, spin is not the only force other than gravity which acts on the projectile. Air resistance can have a significant effect. Considerable experimental evidence is available [6] to indicate that satisfactory models of air resistance are dependent on either the first or second power of the velocity of the ball. For the size and speed of the balls usually involved in sport situations, the quadratic model for the air resistance force $\mathbf{F}_D$ is found to be appropriate:

$$\mathbf{F}_D = -\tfrac{1}{2}\rho A C_D |\mathbf{V}|^2 \hat{\mathbf{V}}$$

where ρ is the air density, A is the cross sectional area presented to the air by the ball, C_D is a dimensional constant called the drag coefficient, $\mathbf{V}$ is the velocity of the ball and $-\hat{\mathbf{V}}$ is a unit vector in the direction opposite to the motion.

For different balls the importance of air resistance can be assessed by considering the magnitude of the coefficient $\rho A C_D/2m$. Assuming constancy of air density and drag coefficient (is it reasonable to assume that C_D has the same value for all balls?) this assessment of relative importance becomes equivalent to using a^2/m as the indicator, since $A = \pi a^2$ for a spherical ball of radius a. The larger the value of a^2/m the greater the effect of air resistance, as can be seen from the equation of motion. Readers are encouraged to draw up a table of values of a^2/m for popular sports such as cricket, golf, squash, tennis and table tennis. The results should indicate that air resistance effects range from being very important in table tennis through to being less so in cricket and golf. A word of caution – don't rule out the air effects in golf and cricket as they play a major role in producing the lift force on a spinning ball, as we saw in the previous model.

Formulate a Mathematical Model

Fig. 6.19 shows a ball, modelled as a sphere of mass m and radius a, at a general point on its trajectory. The only forces acting on the ball are assumed to be its weight and the air resistance force. The equation of motion of the ball is

$$m\mathbf{a} = m\mathbf{g} + \mathbf{F}_D$$

where **a** denotes the acceleration of the ball.

Assuming that the motion of the ball is two-dimensional, then the velocity **V** can be expressed as

$$\mathbf{V} = u\mathbf{i} + v\mathbf{j}$$

so that the equation of motion becomes

$$m\left(\frac{\mathrm{d}u}{\mathrm{d}t}\mathbf{i} + \frac{\mathrm{d}v}{\mathrm{d}t}\mathbf{j}\right) = -mg\mathbf{j} - \frac{\rho A C_D}{2}\left(u^2 + v^2\right) \cdot \frac{(u\mathbf{i} + v\mathbf{j})}{\sqrt{u^2 + v^2}} \tag{6.23}$$

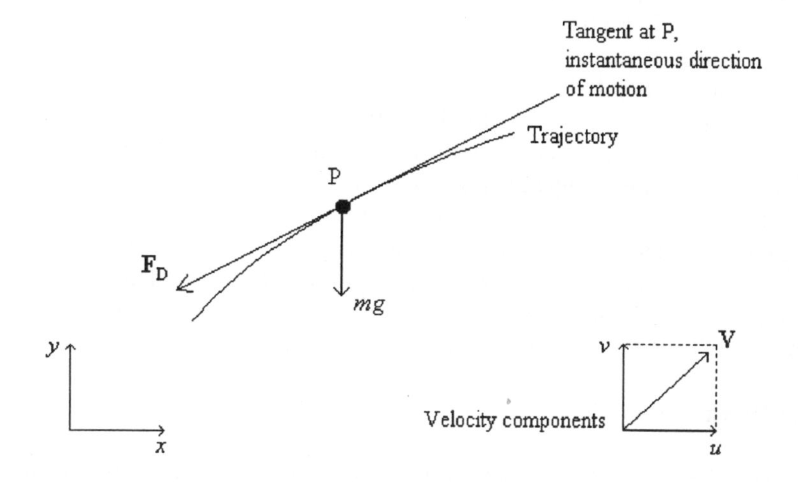

Figure 6.19 Forces acting on a projectile which undergoes two dimensional motion under the action of gravity and air resistance

Recalling that

$$\frac{\mathrm{d}x}{\mathrm{d}t} = u \quad \text{and} \quad \frac{\mathrm{d}y}{\mathrm{d}t} = v$$

and extracting the two component equations of motion from (6.23), the trajectory of the ball is given by the solution of the following system of four coupled first-order ordinary differential equations:

$$\frac{\mathrm{d}x}{\mathrm{d}t} = u$$

$$\frac{dy}{dt} = v$$

$$\frac{du}{dt} = -\frac{\rho A C_D}{2m} u \sqrt{u^2 + v^2}$$

$$\frac{dv}{dt} = -9.81 - \frac{\rho A C_D}{2m} v \sqrt{u^2 + v^2}$$

For purposes of illustration we shall again assume that we are dealing with the baseline drive of a lawn tennis ball (for which $m = 0.058$ kg and radius $a = 0.033$ m). The air density has the value $\rho = 1.25$ kg m^{-3} and, for a new tennis ball, the drag coefficient has the value $C_D = 0.63$, see reference [7], so that $\rho A C_D/2m = 0.0232$. Using this value obtain the Runge–Kutta solution of the system of differential equations using the DERIVE utility ODE_APPR.MTH. The co-ordinates (x,y) required for the trajectory can then be extracted from the solution matrix using the EXTRACT_2_COLUMNS facility. Plotting the (x,y) solution set then gives the trajectory with air resistance being taken into account. For comparison purposes, it is interesting to also plot the trajectory in the absence of air resistance. This can be readily obtained by using the Manage and Substitute commands to amend the terms involving 0.0232 in the above Runge–Kutta expression by replacing it with zero and then re-solving the problem. The resistance free trajectory will of course be parabolic. Plotting both trajectories on the same axes will show you that the air resistance effects are significant in this example.

Further Work

(i) Research the data needed to analyze other tennis strokes and investigate these where spin and/or air resistance is included. To help you get started the following data are representative of a serve in tennis:

$$h = 2.5 \text{ m} ;\ u = 60 \text{ m s}^{-1} \text{ (maximum)} ;\ \alpha = -6°$$

(ii) With the ball moving in the xy plane, combine the two models presented in Section 6.5 and 6.6 to investigate the case of a ball which experiences both air resistance and topspin, or backspin.

(iii) Revise the spin model to deal with spin about the y axis. This results in the ball moving out of its original xy plane of projection so that the problem becomes three-dimensional. The action is akin to a sliced shot in golf or the swing of a cricket ball.

6.7 MODELLING A CAR SUSPENSION SYSTEM

The Problem

Southport is a coastal resort in the north-west of England and, in common with many such resorts, has to overcome problems of traffic congestion within the town and ease of access to and from the town. One particular access road is built upon a raised bank that has marshland on one side, and the beach and more marshland on the other side, and this road has suffered from subsidence over the years as the foundations have settled. The road now has a series of undulations of varying amplitude and frequency over a two-mile stretch, despite attempts to reinforce the foundations.

Driving along this road has become one of the resort's novel attractions. Travelling at around 40 mph. gives an uncomfortable ride with frequent car bounces, which means continual braking (and hence slower inflow/outflow of traffic). The driver can either travel at a slower speed or perhaps try to jump the bumps by travelling at a faster speed. Perhaps of more concern to the individual is the effect on the car suspension system of travelling along this road and the possibility of damage to the car. A model is needed to investigate these effects.

Setting up a Model

Factors likely to influence the investigation can be listed as follows:

- the mass of the car and its contents and the distribution of that mass,
- the car tyres and the wheel axles,
- the shock absorbers,
- the surface profile of the road,
- the contact between the tyres and the road surface,
- the horizontal speed of the car,
- aerodynamic forces and other external forces acting on the car.

We shall first formulate a simple model by making the following assumptions:

- The mass of the car body, including its occupants, engine, etc., can be considered to be uniformly distributed across each of the four wheels, i.e. if M is the mass of the car, a mass $m = M/4$ acts at each wheel.
- The compliance of the tyres is ignored initially.
- The shock absorber is modelled as a simple spring-dashpot system with spring stiffness k and dashpot constant c. Most car suspension systems use springs in the form of a coil or a series of leaves and are usually made of steel, although rubber and plastic composites are possible. The dashpot (or damper, as it is commonly known) is usually a hydraulic device which is effectively a piston moving inside a housing containing fluid.

- For the spring, Hooke's law is assumed to hold, i.e. as the spring extends, the resisting force F is proportional to the spring extension, the constant of proportionality being the spring stiffness k.
- For the dashpot, it is assumed that the resisting force is proportional to the relative velocity of the housing and the piston, the constant of proportionality being the dashpot constant c.
- The road surface profile can be described as a sinusoid and has an equation of the form

$$y = h \sin(\alpha z)$$

where h is the amplitude of the sinusoid, y is the vertical displacement of the wheel due to the road surface measured with respect to some fixed horizontal datum line and z is horizontal displacement (see Fig. 6.20).

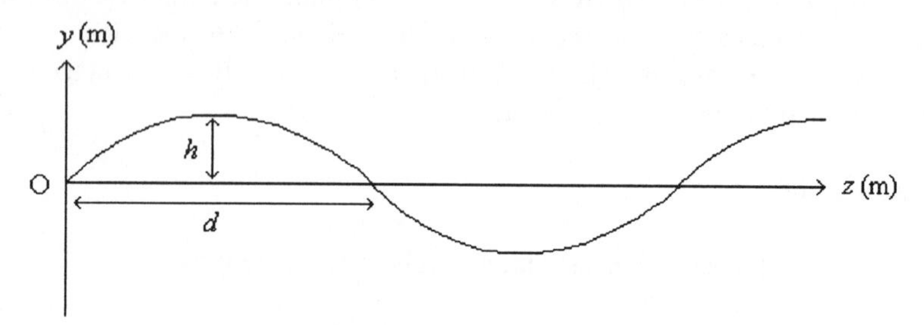

Figure 6.20 Road surface profile

If x is the vertical displacement of the car body above its equilibrium position, then this model is as illustrated in Fig. 6.21.

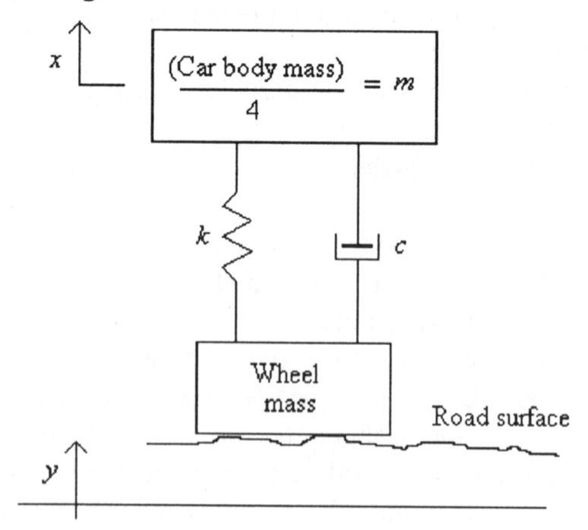

Figure 6.21 Simple car body model

Formulate a Mathematical Model

To derive the mathematical equations describing the motion, we apply Newton's second law of motion in a vertical direction. Assuming both the car body and the wheels are moving vertically upwards with the spring and damper (dashpot) both extended, we have:

Extension of the spring $= x-y$

so,

Spring force which resists motion, $F_s = k(x-y)$

Resistance to motion due to the damper is given by $F_d = c(\dot{x} - \dot{y})$, where the dot denotes differentiation with respect to time t. Hence, applying Newton's second law of motion vertically upwards, we have (see Fig. 6.22)

$$\begin{aligned} m\ddot{x} &= -F_s - F_d \\ &= -k(x-y) - c(\dot{x}-\dot{y}) \end{aligned}$$

which rearranges to give the differential equation

$$m\ddot{x} + c\dot{x} + kx = ky + c\dot{y} \tag{6.24}$$

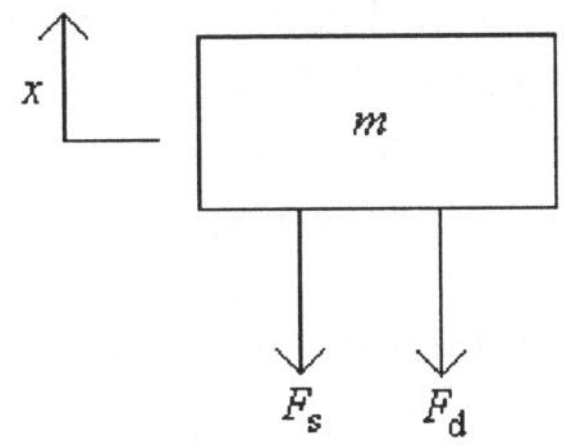

Figure 6.22 Car body force diagram

Equation (6.24) is a second-order (since the highest-order derivative present is two), linear (since x and derivatives of x occur linearly), differential equation with constant coefficients. This type of differential equation commonly occurs in the study of control systems and is often expressed in the form

$$\ddot{x} + 2\xi\omega_n\dot{x} + \omega_n^2 x = r(t)$$

where $r(t)$ is the applied forcing term and ξ and ω_n are constants (see, for example, [8]).

The mathematical formulation is not yet complete. In order to be able to solve equation (6.24) the right-hand side needs to be expressed as a function of time t. This can be done if we further assume that the car is travelling with an average horizontal speed V and hence that $z = Vt$. Making this assumption, the form for y can be conveniently written as

$$y = h\sin(\pi Vt / d)$$

where h and d are as shown in Fig. 6.20 and equation (6.24) becomes

$$m\ddot{x} + c\dot{x} + kx = kh\sin(\pi Vt / d) + ch(\pi V / d)\cos(\pi Vt / d) \qquad (6.25)$$

To find a particular solution of equation (6.25), two conditions need to be specified. We shall suppose that, initially, the system is in equilibrium and hence that

$$x = \dot{x} = 0 \qquad \text{at } t = 0 \qquad (6.26)$$

Thus our mathematical problem is to solve equation (6.25) subject to the initial conditions (6.26).

Solution of the Mathematical Model

Equation (6.25) can be solved by applying standard mathematical techniques such as the complementary function-particular integral method or by taking Laplace transforms. Here we shall use DERIVE to assist in obtaining a solution by finding a complementary function and adding a particular integral to this.

Before proceeding with the solution, it is convenient to consider specific values for the parameters in equation (6.25). Values representative of a Ford Scorpio car are

$$m = 275 \text{ kg}, \qquad k = 25\,000 \text{ N m}^{-1}, \qquad c = 1200 \text{ Ns m}^{-1}$$

while the authors have found the following traffic speed and undulation parameters to be representative for the road described earlier:

$$V = 14 \text{ m s}^{-1} \ (\triangleq 30 \text{ mph}), \qquad d = 2 \text{ m}, \qquad h = 0.1 \text{ m}$$

We are now in a position to solve the problem using DERIVE. Details are presented in Fig. 6.23.

Firstly, the differential equation is established, with the values for m, c, k, v, d and h being Authored, expressions 1–6 (This could be avoided and the differential equation simply Authored with the specific values entered. However, if it is desired to see the effects of varying parameter values, it is useful to proceed as above.) The road surface profile y can now be Authored as expression 7.

Finally, the car body displacement $x(t)$ is declared as expression 8.

```
1:   m := 275

2:   c := 1200

3:   k := 25000

4:   v := 14

5:   d := 2

6:   h := 0.1

                 ┌ π v t ┐
7:   y := h SIN │───────│
                 └   d   ┘

8:   X(t) :=

       ┌d ┐2            ┌d ┐1
9:   m │──│  X(t) + c │──│  X(t) + k X(t)
       └dt┘            └dt┘

                ┌d ┐1
10:  k y + c │──│  y
                └dt┘

       ┌d ┐2            ┌d ┐1                          ┌d ┐1
11:  m │──│  X(t) + c │──│  X(t) + k X(t) = k y + c │──│  y
       └dt┘            └dt┘                          └dt┘

         ┌d ┐2                d
12:  275 │──│  X(t) + 1200 ── X(t) + 25000 X(t) = 840 π COS(7 π t) + 2500 SIN(7
         └dt┘                dt

                w t
13:  X(t) := ê

      t w          2
14:  ê     (275 w   + 1200 w + 25000)

              24     2 √2606 î
15:  w = - ──── - ─────────
              11        11

              24     2 √2606 î
16:  w = - ──── + ─────────
              11        11

17:  X(t) := a COS(7 π t) + b SIN(7 π t)

                 2                                                     2
18:  - 25 (539 π  a - 336 π b - 1000 a) COS(7 π t) - 25 (539 π  b + 336 π a - 10

                   2                                            2
19:  [- 25 (539 π  a - 336 π b - 1000 a) = 840 π, - 25 (539 π  b + 336 π a - 10

                                3
      ┌                  90552 π                                4 (125000 - 53263
20:   │a = - ─────────────────────────────────, b = ──────────────────────
      │                   4               2                       4             2
      └     5 (290521 π  - 965104 π  + 1000000)      5 (290521 π  - 965104 π

      ┌        387            543  ┐
21:   │a = - ─────, b = - ─────│
      └       13628          33497 ┘
═══════════════════════════════════════════════════════════════════════════════
COMMAND: Author Build Calculus Declare Expand Factor Help Jump soLve Manage
         Options Plot Quit Remove Simplify Transfer moVe Window approX
Enter option
User                    A:CAR1.MTH          Free:86%  Ins           Derive Algebra
```

Figure 6.23 Solution of the car suspension problem

```
           - 24 t / 11  ┌      ┌ 2 √2026 t ┐        ┌ 2 √2026 t ┐┐
22:  ê                  │c1 COS│───────────│ + c2 SIN│───────────││ - 0.0283974 COS(7
                        └      └    11     ┘        └    11     ┘┘

           - 24 0 / 11  ┌      ┌ 2 √2026 0 ┐        ┌ 2 √2026 0 ┐┐
23:  ê                  │c1 COS│───────────│ + c2 SIN│───────────││ - 0.0283974 COS(7
                        └      └    11     ┘        └    11     ┘┘

            141987
24:  c1 - ───────
           5000000

     d   ┌ - 24 t / 11  ┌      ┌ 2 √2026 t ┐        ┌ 2 √2026 t ┐┐
25:  ──  │ê             │c1 COS│───────────│ + c2 SIN│───────────││ - 0.0283974 CO
     dt  └              └      └    11     ┘        └    11     ┘┘

                       ┌                                   ┌ 2 √2026 t ┐
                       │ 2 √2 (6 √2 c1 - √1013 c2) COS│───────────│        2 √2 (√1013
26:       - 24 t / 11  │                                   └    11     ┘
     - ê               │───────────────────────────────────────────────── + ───────────
                       └                        11

                       ┌                                   ┌ 2 √2026 0 ┐
                       │ 2 √2 (6 √2 c1 - √1013 c2) COS│───────────│        2 √2 (√1013
27:       - 24 0 / 11  │                                   └    11     ┘
     - ê               │───────────────────────────────────────────────── + ───────────
                       └                        11

       141841 π     2 √2 (6 √2 c1 - √1013 c2)
28:  - ──────── - ─────────────────────────────
       1250000                 11

     ┌        141987           141841 π     2 √2 (6 √2 c1 - √1013 c2)      ┐
29:  │c1 - ─────── = 0, - ──────── - ───────────────────────────── = 0│
     └       5000000           1250000                 11                    ┘

     ┌       141987           1560251 √2026 π     425961 √2026 ┐
30:  │c1 = ───────, c2 = ──────────────── + ─────────────│
     └      5000000             5065000000          2532500000  ┘

31:  [c1 = 0.0283973, c2 = 0.0511304]

COMMAND: Author Build Calculus Declare Expand Factor Help Jump soLve Manage
         Options Plot Quit Remove Simplify Transfer moVe Window approX
Compute time: 0.1 seconds
Approx(30)              A:CAR1.MTH              Free:91%              Derive Algebra
```

Figure 6.23 Solution of the car suspension problem (concluded)

The differential equation (6.25) is now Authored and Simplified. This is done by using the DERIVE function dif(u,x,n) so that d^2x/dt^2 is entered as dif($x(t)$,t,2) and dx/dt as dif($x(t)$,t,l), or simply dif($x(t)$).

Rather than Author a lengthy equation like (6.25) all at once, it is easier to Author the left- and right-hand sides separately (as expressions 9 and 10) and then equate by Authoring expression 9 = expression 10 to give expression 11. This is then Simplified (expression 12).

Adopting the complementary function–particular integral method, we shall first find the complementary function.

Author the expression $x(t) = e^{wt}$, (expression 13) .

Simplify and soLve expression 9 to obtain the roots of the auxiliary equation (see expressions 15 and 16). Hence the complementary function is

$$e^{-24t/11}\left[c_1\cos\left(2\sqrt{2606}\,t/11\right)+c_2\sin\left(2\sqrt{2606}\,t/11\right)\right]$$

where c_1 and c_2 are arbitrary constants.

To find the particular integral, we adopt a trial function of the form

$$x(t) = a\cos(7\pi t) + b\sin(7\pi t)$$

(i.e. a similar form to the right-hand side of the differential equation (6.25)). This is Authored (expression 17), expression 9 is Simplified (expression 18) and the coefficients of the cosine and sine terms compared with the right-hand side of expression 12. The resulting pair of algebraic equations is then Authored (expression 19) and soLved to give a solution for a and b which is then approXimated for convenience (expression 21).

The general solution is then the sum of the complementary function and the particular integral

$$x(t) = e^{-24t/11}\left[c_1\cos\left(2\sqrt{2026}\,t/11\right)+c_2\sin\left(2\sqrt{2026}\,t/11\right)\right] \\ -0.028\,397\,4\cos(7\pi t)-0.016\,210\,4\sin(7\pi t) \qquad (6.27)$$

The particular solution satisfying conditions (6.26) is then found by Authoring equation (6.27) and using Manage, Substitute, for $t = 0$ (remember to declare real variables c_1 and c_2 first). Simplifying gives rise to an expression for c_1 (expression 24).

Similarly, Calculus, Differentiate expression 22 and Manage, Substitute $t = 0$. This gives rise to an expression involving c_1 and c_2 (see expression 28).

SoLve expressions 24 and 28 as a pair of simultaneous equations to give

$c_1 = 0.0284$ and $c_2 = 0.0511$ (expressions 29–31)

The above has seen DERIVE used to assist with the mathematical manipulations needed to apply a known analytical method of solution.

DERIVE has a set of utility functions that can assist us more readily with this problem. In particular, select Help and move through the pages until section 9.7 is located, where you will find details of the function

DSOLVE2_IV (p,q,r,x,x_0,y_0,v_0)

which solves the second-order differential equation

$$y'' + p(x)y' + q(x)y = r(x)$$

with initial conditions

$$y = y_0 \qquad \text{and} \qquad dy/dx = v_0 \qquad \text{at} \qquad x = x_0$$

Thus, for our problem,

$$p = c/m,\ q = k/m,\ r = \frac{ky + c\dot{y}}{m},\quad x_0 = y_0 = v_0 = 0$$

the dependent variable is x and the independent variable is t.

To use the above DERIVE function first load the utility file ODE2.MTH and then Author the expression shown as number 32 in Fig. 6.24. Expression 33 is the Simplified form and expression 34 the approXimate form. Note that expression 34 looks somewhat different with respect to the particular integral compared with that naturally occurring via the complementary function–particular integral method used earlier. Trigonometric manipulation will show that the two forms are the same (this is left as an exercise for the reader), as will a plot of the two forms.

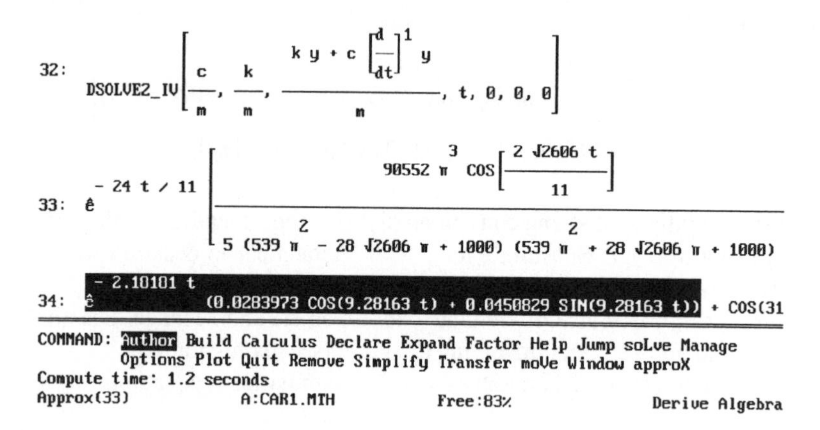

Figure 6.24 Solution of the car suspension problem using DERIVE's utility functions

Interpretation and Validation

In order to validate the results of this model, the engineers would probably carry out some experimental measurements of $x(t)$ using transducers or similar measuring devices and compare them with $x(t)$, as predicted by expression 34.

Our assessment of its validity is based on different criteria. We shall examine how the model predicts the following:

(i) The suspension travel limits – for the Ford Scorpio the extent of travel of the spring must lie within $\pm$ 0.1 m. Our model should therefore predict this.

(ii) Passenger discomfort – discomfort is perceived if the magnitude of the car body's vertical acceleration exceeds some limit and a design criterion often adopted is that this upper limit is of the order of 0.6 g.

Thus (i) and (ii) imply

$$|x-y| \leq 0.1 \quad \text{for all values of time } t$$

$$|\ddot{x}| \leq 0.6g \quad \text{for all values of time } t \qquad (g = 9.8 \text{ m s}^{-2})$$

DERIVE's graph plotting facility can be used to advantage to assess the extent to which these conditions are met.

From expression 34, expressions for $x-y$ and $\ddot{x}$, (dif(#34,t,2)), were Authored and Plotted (see Figs. 6.25 and 6.26 respectively). The reader is encouraged to derive and plot these expressions and experiment with the Scale and Centre facilities. The cross-hair can then be used to estimate function extrema.

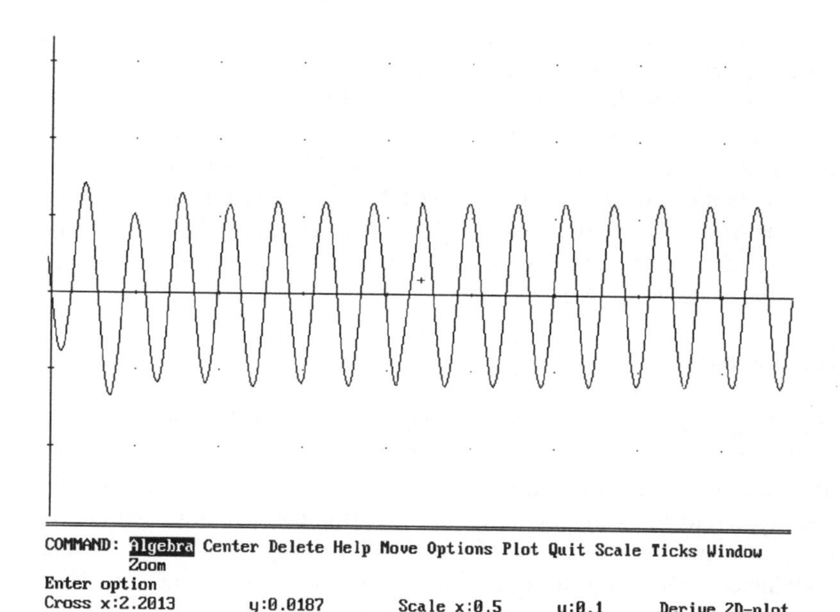

Figure 6.25 Plot of displacement $x-y$ as a function of time

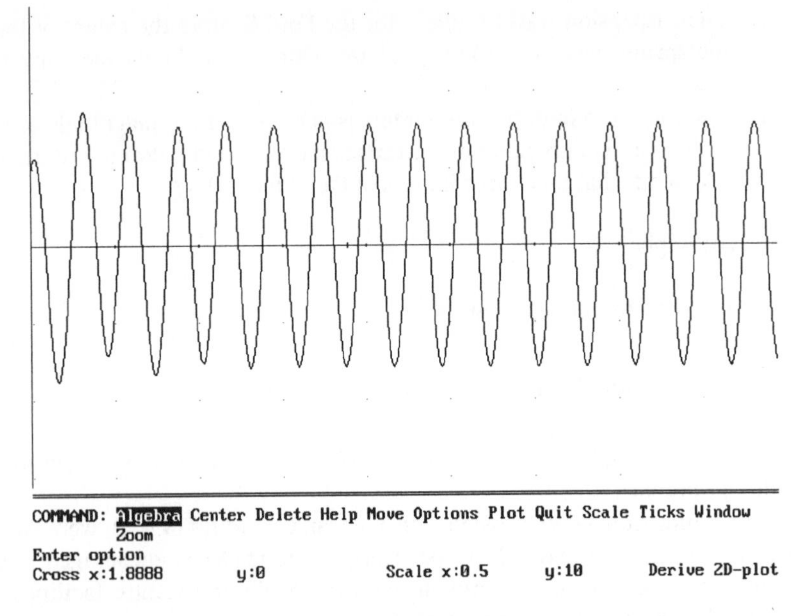

Figure 6.26 Plot of acceleration $\ddot{x}$ as a function of time

The graphs indicate that both bounds implied by (i) and (ii) are exceeded for some value of time t. Fig. (6.26) and the authors' own experience of travelling along the road in question (although admittedly not in a Ford Scorpio) would suggest that the car ride is indeed uncomfortable and therefore that this should be predicted. The failure to predict bounds of ± 0.1 for suspension travel is a weakness of the model.

To summarize, the proposed model has the advantage of having a readily available mathematical solution which will enable a system designer to investigate the effects of parameter changes. The predicted behaviour fails to meet the design criteria and hence the model needs some revision.

Model Refinement

How can the previous model be improved bearing in mind that the more complex the model, the more likely it is that a mathematical solution will be difficult to find? Assuming for the moment that the quoted values for k and c are appropriate, there is no particular reason to doubt the usefulness of the spring-dashpot suspension model. Most of the other assumptions appear to be reasonable except, perhaps, for the omission of tyre compliance. It would seem sensible to try and include wheel and axle behaviour in our model, using further spring-dashpot components. Thus our refined model is illustrated in Fig. 6.27.

The mathematical formulation proceeds in a manner similar to before. The forces acting on each mass are illustrated in Fig. 6.28 and in each case Newton's second law of motion is applied vertically upwards with the same assumptions as before.

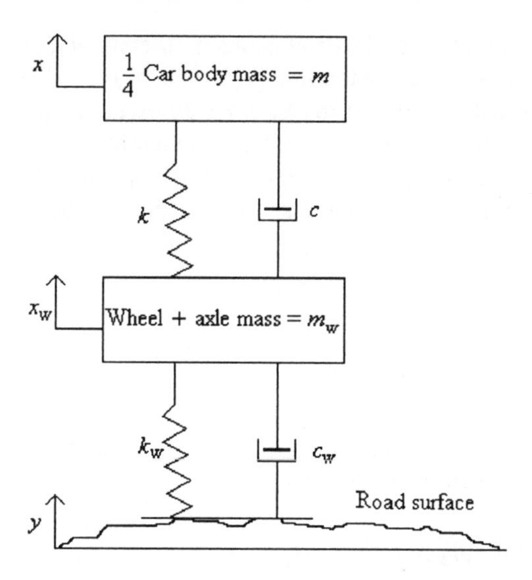

Figure 6.27 Refined car body model

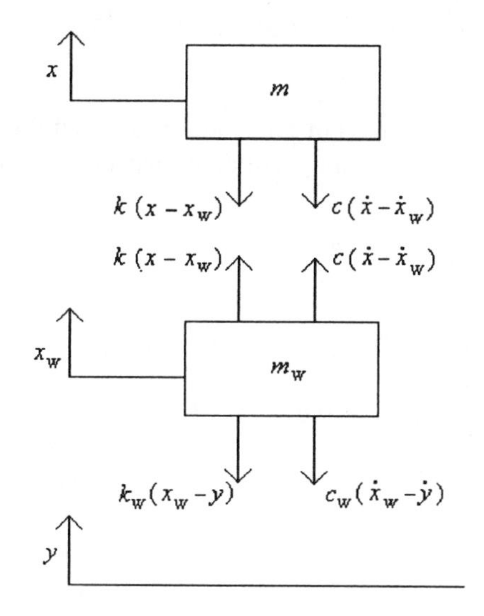

Figure 6.28 Forces diagram for the refined car body model

The equations of motion then become

$$m\ddot{x} \quad = \quad -k(x-x_w) - c(\dot{x}-\dot{x}_w) \tag{6.28}$$

$$m_w\ddot{x}_w = k(x-x_w) - c(\dot{x}-\dot{x}_w) - k_w(x_w-y) - c_w(\dot{x}_w-\dot{y}) \tag{6.29}$$

Equations (6.28) and (6.29) are a pair of coupled differential equations and they need to be solved simultaneously to determine $x(t)$ and $x_w(t)$. This is not particularly straightforward and will result in a fourth-order differential equation if elimination of x or x_w is attempted, with little hope of a closed-form mathematical solution.

A more useful way to proceed is to re-write equations (6.28) and (6.29) as a set of four first-order differential equations. To do this we introduce variables u, v as follows:

Let

$$\dot{x} = u \tag{6.30}$$

$$\dot{x}_w = v \tag{6.31}$$

then (6.28) and (6.29) can be written as

$$\dot{u} = \frac{1}{m}\{-k(x-x_w) - c(u-v)\} \tag{6.32}$$

$$\dot{v} = \frac{1}{m_w}\{k(x-x_w) + c(u-v) - k_w(x_w-y) - c_w(v-\dot{y})\} \tag{6.33}$$

Equations (6.30)–(6.33) constitute our set of four first-order differential equations in the four variables x, x_w, u, v. These can be written in matrix form as

$$\dot{\mathbf{x}} = A\mathbf{x} + \mathbf{b}$$

where $\mathbf{x}$ is the vector $\begin{bmatrix} x \\ x_w \\ u \\ v \end{bmatrix}$ and $\mathbf{b} = \begin{bmatrix} 0 \\ 0 \\ 0 \\ ky + c_w\dot{y} \end{bmatrix}$

$$A = \begin{bmatrix} 0 & 0 & 1 & 0 \\ 0 & 0 & 0 & 1 \\ -k/m & k/m & -c/m & c/m \\ k/m_w & \dfrac{-(k+k_w)}{m_w} & c/m_w & \dfrac{-(c+c_w)}{m_w} \end{bmatrix}$$

Thus our mathematical problem is now to solve the system of first-order differential equations (6.30)–(6.33) subject to the conditions

$$x = x_w = u = v = 0 \qquad \text{at} \quad t = 0$$

(i.e. the system starts from the equilibrium position).

Once again, the DERIVE utility file can help us by providing the Runge–Kutta function RK which has been used earlier in the solution of the foxes and rabbits and the drug transport problems.

Remembering to Transfer, Load the utility file ODE_APPR.MTH first, and once values for m, m_w and the other parameters have been Declared, the *RK* function is Authored as the expression

$$\begin{aligned}\text{RK}\Big(\big[u, v, (-k(x-x_w) - c(u-v))/m, (k(x-x_w) + c(u-v) - k_w(x_w - y)\\ -c_w(v - ydot))/m_w\big], [t, x, x_w, u, v], [0,0,0,0,0], 0.25, 60\Big)\end{aligned} \tag{6.34}$$

The parameters involved here are those for the Ford Scorpio, namely,

$$m = 275 \text{ kg}, \qquad m_w = 45 \text{ kg}, \qquad k = 25\,000 \text{ N m}^{-1}$$

$$k_w = 250\,000 \text{ N m}^{-1}, \qquad c = 1200 \text{ Ns m}^{-1}, \qquad c_w = 140 \text{ Ns m}^{-1}$$

We shall assume the same road surface profile (y) as before, ydot being its time derivative $\mathrm{d}y/\mathrm{d}t$.

The last two parameters of the RK function are the step-size and the number of steps. The values specified in equation (6.34) are for illustrative purposes only (and in fact are the values used in the foxes, rabbits problem). Bear in mind that the Runge–Kutta method is an approximate solution method and that the smaller the step-size the better. On the other hand, the greater the number of steps specified, the longer the computation time. As

$$\textit{Step-size} \times \textit{number of steps} = \textit{range of time } t$$

then some compromise has to be reached by trial and error.

Authors' note. The authors have experienced some difficulty with DERIVE when attempting to approXimate equation (6.34). The declaration of variables and constant values prior to the Authoring of (6.34) and the choice of variable names u and v resulted in output that contained variable expressions rather than numerical values. In some cases the 'memory full' message was displayed. To overcome these difficulties an explicit form of equation (6.34) was Authored as

$$\text{RK}\Big(\Big[\,p, q\,, (-25\,000(x-z) - 1200(p-q))\,/\,275, \big(25\,000(x-z) + 1200\,(p-q) - 250\,000\,(z - 0.1\sin(7\pi t)) - 140\,(q - 0.7\pi\cos(7\pi t))\big)\,/\,45\,\Big], [\,t, x, z, p, q\,], [\,0, 0, 0, 0, 0\,], 0.25, 60\,\Big)$$

The reader should observe that when approXimated (remember to use approximate precision), the solution vector contains large numerical values, even at small values of t. Such values are clearly inappropriate for this car suspension problem and are evidence that the step-size is too large and numerical instability is occurring. The authors have found by experimentation that a step-size of 0.01 appears to give reasonable values, but the reader should conduct similar experiments to confirm this and have confidence in the solution. A simple experimental procedure is to keep halving the step-size (use Manage Substitute) and re-compute the solution until there is no significant change in the solution.

Now that the refined model has been established, the reader is encouraged to use DERIVE to complete the solution process and attempt to compare models in the following way:

(a) Solve the system equations for x, x_w, $\dot{x}$ and $\dot{x}_w$ using the Runge–Kutta function for as large a value of t as possible.
(b) Use the EXTRACT_2_COLUMNS file function to extract appropriate columns of data to plot x against t and x_w against t.
(c) Use DERIVE utilities to obtain expressions for $|x - x_w|$ and $|\ddot{x}|$, the suspension travel limit and passenger discomfort measures considered earlier. (Hint: you may need to look at the utility file function 9.3.)
(d) Plot $x - x_w$. and $\ddot{x}$ against t and observe whether the limits quoted earlier are exceeded or not.
(e) Compare the predictions for the two models critically.

Discussion

The modelling of car suspensions continues to develop as they become more sophisticated and computer software is developed to solve the increasingly complex models. The two models considered here serve to illustrate the modelling process and although they are still based on linear ordinary differential equations, they are nevertheless sufficiently complex to stretch DERIVE's capabilities. None the less, it is worthwhile considering how other models could be developed, even though the reader should be aware that increased complexity does not necessarily imply a more useful model.

One obvious extension to our previous models is to divide the car into two rather than four and consider the engine, seats, front wheels and back wheels as units of the model (see Fig. 6.29). The modelling process and derivation of the mathematical equations can proceed exactly as above but there are of course more of them. (How many exactly?) Data would be needed for representative values of all the masses, springs and dampers in the model.

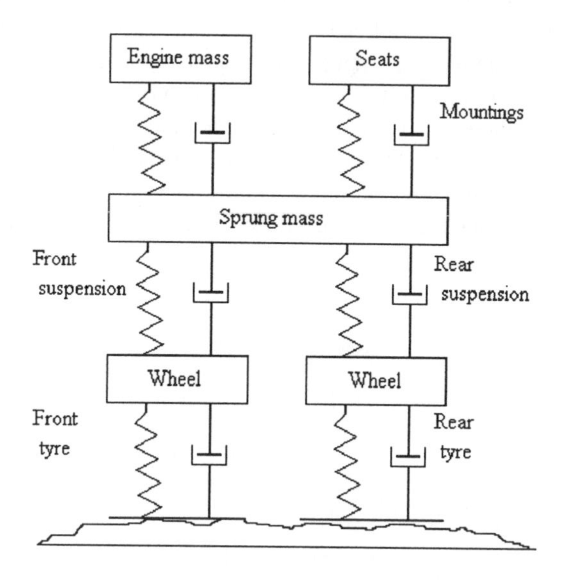

Figure 6.29 Extended car body model

The next point of issue concerns the spring and damper models themselves. These components have to satisfy the demands of passenger comfort and vehicle handling for a wide variety of road surface profiles and load conditions. It is perhaps optimistic to think that a fixed value for the spring constant or the damping constant will suffice to maximize performance under all conditions. For the damper in particular, a low setting is best for a comfortable ride, while a high setting is required to improve vehicle handling so a variable damper would seem preferable and these are commercially available.

This ability to vary the damping force is a design feature that can be incorporated into our system model. For the Ford Scorpio, c = 1200 Ns m^{-1} is in fact an average figure and c ranges from 600 Ns m^{-1} when the damper is compressed to 1800 Ns m^{-1} when the damper is extending.

For the model developed first, we could introduce the following condition:

if $\dot{x} - \dot{y} < 0$ then $c = 600$

else $c = 1800$

In principle, this condition can be incorporated into the mathematical solution by solving the second-order differential equation starting with $c = 600$ (going uphill at the start), solving for $x(t)$ and then finding when $\dot{x} - \dot{y} = 0$. For this value of time t, switch to $c = 1800$ and re-solve the differential equation with the starting conditions for $c = 1800$ matching the end conditions for $c = 600$. And so on!

In practice, the authors have found that DERIVE has difficulty in solving $\dot{x} - \dot{y} = 0$ and the choice is either to use an approximate method like Newton-Raphson or to estimate values from the plotted graphs – the reader is encouraged to try it out!

Finally in this discussion, we note that the models considered here are so-called PASSIVE suspension models. One hears considerable talk of ACTIVE suspension systems, particularly in Formula One motor racing, where in essence, the forces acting on the body mass are computer-monitored and the suspension parameters are varied according to the instantaneous values of these forces. Such a model is illustrated schematically in Fig. 6.30.

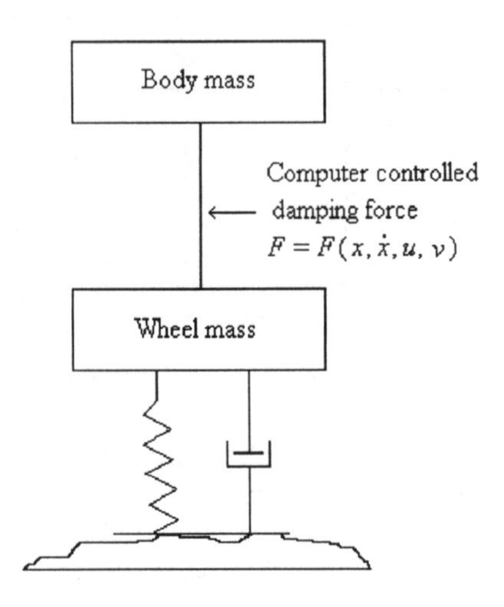

Figure 6.30 Active suspension model

Further Work

1. Try solving the simple second-order differential equation model with $c = 600$ throughout and then again with $c = 1800$ to get a feel for the effect of changing the dashpot constant. In particular, note the prediction for suspension travel limits and passenger discomfort level described earlier. Do these values give rise to more confidence in the suitability of the model or not?
2. As the above discussion suggests, try using the graph plotting facility of DERIVE to estimate the first time when $\dot{x} - \dot{y} = 0$ and hence c changes from 600 to 1800 for a variable damper model. Again consider whether the results give you more or less confidence in the model.

Many variations are possible for y, the road surface profile. Two suggested exercises here are as follows.

3. Use the simple model to investigate the effect of travelling over the bumps at 45 mph, which is approximately 20 m s^{-1}, i.e. $y = 0.1 \sin(10\pi t)$.
4. Model the situation when a car moves over a series of 'sleeping policemen', illustrated by Fig. 6.31.

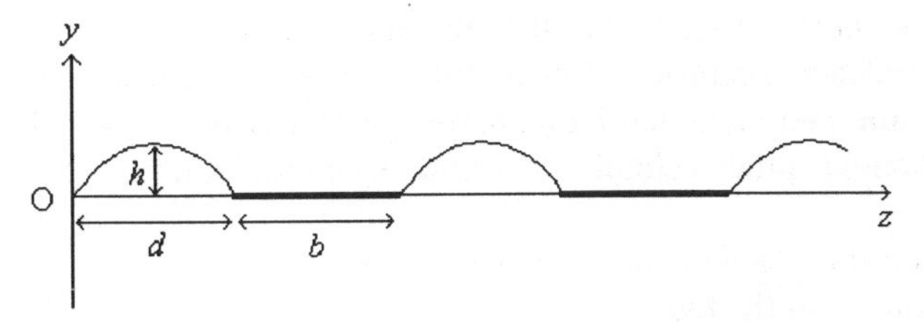

Figure 6.31 'Sleeping policemen' road profile

Hint: y is a periodic function and can be represented by a Fourier series. Considerable use can be made of the DERIVE special function CHI (a,x,b) and the utility function FOURIER (see Help utility, section 9.5). The mathematical formulation of the Fourier series can be found, for example, in [9]. You are recommended to take only a few terms of the series.

Once y is established as a Fourier series then DSOLVE2_IV can be used as before to determine a solution for $x(t)$.

Note: You may wish to consider here the possibility that the driver races along the flat parts and brakes hard for the bumps. How can this be incorporated into the model?

6.8 A STRATEGY FOR LEARNING WITH MINIMUM EFFORT

The Problem

The goal of many students is to maximize the amount of knowledge gained from a course (or at least gain enough to pass the examination!) while at the same time minimizing the intellectual effort expended in achieving it. Your own experience will tell you that different study strategies exist. Develop a mathematical model which could be used to establish a work schedule which would permit a student to achieve the above goal.

Setting up a Model

As is usual when attempting to model problems for which a specific mathematical approach does not immediately suggest itself, we begin by constructing a features list which will serve to focus our ideas as to what might be important in analysing the problem.

Since we all have some experience of studying we can use it to assist us in producing a more manageable pruned features list. Our experience is that we do forget some of the information previously learnt and that we can only work at a high level of concentration for a limited time. Furthermore, the rate at which we can learn new material depends to some extent on its nature: a set of discrete facts (history dates, for example) poses more difficulty for the authors than a mathematical proof (which exhibits a logical development). Finally, at the end of the course we need to have retained a certain minimum level of knowledge sufficient to pass the examination! These initial thoughts prove helpful in constructing an initial features list, such as:

- interest in the topic to be studied,
- aptitude for the topic,
- duration of the course,
- time for which the student can concentrate,
- time available for study,
- how much knowledge is retained from one study session to the next,
- how much knowledge the student possesses to begin with,
- whether the topic is composed of discrete facts or develops logically,
- need to pass the examination,
- and so on.

Your features list may well differ from this. Don't worry. As we have mentioned previously, there is no unique answer when modelling. The acid test is whether your solution will stand the 'Do I believe this result?' test!

Reflection on our studying experiences led to the following pruned list (remember the benefits of dialogue and ranking if you have difficulty in achieving any effective pruning):

- level of knowledge possessed at start of course,
- level of knowledge needed to pass the examination,
- rate at which the students forget knowledge previously learnt,
- duration of the course,
- study pattern.

To progress from here we must make some modelling assumptions, from which the mathematical model will develop, and of course we must define the variables we shall be using.

Denote the student's knowledge at time t by $K(t)$ and the study effort expended by $E(t)$. We shall then assume the following:

- Students possess an initial knowledge level (K_0) and are interested in the topic to be studied.

- Students' increase in knowledge is dependent on their current knowledge level and the study effort expended.
- Students forget a fraction (m) of what they already know (m is a positive constant).
- At the end of the course a certain minimum level of knowledge (K_e) is needed in order to pass the examination.

Formulate a Mathematical Model

The first thing to decide is whether a discrete or continuous model is appropriate. Since student learning is (or at least should be!) a continuous process, then it seems reasonable to opt for a continuous model.

The rate at which knowledge is acquired is thus represented by the derivative $\mathrm{d}K/\mathrm{d}t$ and the rate at which knowledge is lost due to poor memory, etc., is $-mK$ (the minus sign ensuring a loss). Since the growth of the knowledge base is assumed to be dependent on both K and E, then we can model the acquisition of knowledge by

$$\lambda K^a E^b$$

where λ is a constant of proportionality (assumed to be related to the individual student's intelligence) and a and b are constants, whose values are as yet unknown.

The mathematical model of knowledge growth can then be expressed as

$$\frac{\mathrm{d}K}{\mathrm{d}t} = \lambda K^a E^b - mK \tag{6.35}$$

The students' objective is to minimize the total learning effort which is represented by the integral

$$\textit{Total learning effort} = \int_0^T E \, \mathrm{d}t \tag{6.36}$$

where T is the duration of the course.

Any attempt to solve equation (6.35) for $K(t)$ will require the formulation of a model for $E(t)$, the study effort. Several possibilities suggest themselves (no doubt you can think of others too):

(a) Study at a steady rate, i.e. $E = E_0$, a constant.
(b) Study as hard as you can for as long as you can (unlimited effort).
(c) Study for a fixed period each day (so that an upper limit is set for the study effort).

We begin by developing a solution based on model (a). With $E = E_0$, equation (6.35) becomes

$$\frac{dK}{dt} + mK = \lambda K^a E_0^b \tag{6.37}$$

Solution of the Mathematical Model

Equation (6.37) is a first-order differential equation so one is led naturally to examine the functions available within the utility file ODE1.MTH. Of these our equation fits the pattern associated with DSOLVE1. However, when solution is attempted using this function, DERIVE cannot produce the solution. Comparing equation (6.37) with the structure of the differential equations associated with the other functions in DSOLVE1 indicates only one other possibility, INTEGRATING_FACTOR. Alas, this also fails to give a solution.

Nor can we resort to numerical methods since no values are available for the various parameters m, λ, a, b, and E_0. It is thus apparent that DERIVE is unable to provide any significant assistance in the solution of equation (6.37). There is a valuable lesson to be learnt here, namely that mathematical software cannot always provide solutions to every problem. So what can we do now?

Perhaps it is possible to effect a substitution of some sort on equation (6.35) so that the solution is attainable, maybe with the assistance of DERIVE. Although equation (6.37) is non-linear (due to the presence of the K^a term, assuming $a \neq 1$), the left-hand side does at least have the form of an integrating factor type of equation. Therefore, it might be worthwhile multiplying (6.37) through by the integrating factor suggested by the left-hand side, namely e^{mt}, to produce

$$e^{mt} \frac{dK}{dt} + me^{mt} K = \lambda K^a e^{mt} \cdot E_0^b$$

i.e.

$$\frac{d}{dt}(Ke^{mt}) = \lambda K^a e^{mt} \cdot E_0^b$$

(Note the modelling approach that if something seems a reasonable idea then try it.) It is now a simple step to re-write the right-hand side of this equation so that the independent variable may be considered to be Ke^{mt}; thus

$$\frac{d}{dt}(Ke^{mt}) = \lambda(Ke^{mt})^a \cdot e^{-amt} e^{mt} E_0^b \tag{6.38}$$

This is still non-linear (the power of a again) but we can at least make it linear via the following substitution (separation of variables is also possible in terms of the independent variable Ke^{mt}. Why not explore this avenue?):

$$Z = (Ke^{mt})^{1-a} \tag{6.39}$$

so that

$$\frac{dZ}{dt} = (1-a)(Ke^{mt})^{-a} \cdot \frac{d}{dt}(Ke^{mt})$$

and therefore on substitution equation (6.38) becomes

$$\frac{dZ}{dt} = \lambda q \, e^{qmt} \cdot E_0^b \quad ; \quad q = 1-a$$

which is readily integrable using DERIVE to give

$$Z = \frac{\lambda}{m} E_0^b \cdot e^{qmt} + A \tag{6.40}$$

where A is an arbitrary constant of integration (beware – DERIVE does not display the arbitrary constant!).

The substitution used might seem to have been plucked from thin air and it should be noted that it represents the distillation of several attempts by the authors to simplify the form of equation (6.38). Basically, what we were looking for was an expression which after one differentiation with respect to time would produce terms of both $(Ke^{mt})^a$ or $(Ke^{mt})^{-a}$ and $d(Ke^{mt})/dt$ to help simplify equation (6.38). As you can see, the Z substitution presented achieves the desired aims but we had several fruitless attempts before reaching it.

We next formulate the initial knowledge base assumption mathematically as

$$Z = K_0^q \quad \text{at} \quad t = 0$$

and apply it to equation (6.40) to find A so that equation (6.40) becomes, after substitution for A,

$$Z(t) = \frac{\lambda E_0^b}{m}(e^{qmt} - 1) + K_0^q \tag{6.41}$$

Also, by the end of the course ($t = T$) the students must have acquired a certain level of knowledge (K_e) in order to pass the examination, i.e.

$$Z = (K_e \mathrm{e}^{mt})^q \quad \text{at } t = T \tag{6.42}$$

Applying this condition to equation (6.41) and solving for E_0^b gives

$$E_0^b = \frac{m\left(K_e^q \mathrm{e}^{qmT} - K_0^q\right)}{\lambda\left(\mathrm{e}^{qmT} - 1\right)} \tag{6.43}$$

From equation (6.41) the knowledge acquired by time *t* is given by

$$K(t)^q = \left(\frac{\lambda E_0^b}{m}\left(\mathrm{e}^{qmt} - 1\right) + K_0^q\right)\mathrm{e}^{-qmt} = \frac{\lambda E_0^b}{m} + \left(K_0^q - \frac{\lambda E_0^b}{m}\right)\mathrm{e}^{-qmt}$$

Interpretation

Obviously, we want the knowledge level to increase with time and hence in order to counteract the negative exponential we need

$$K_0^q - \frac{\lambda E_0^b}{m} \le 0$$

i.e.

$$E_0 \ge \left(\frac{mK_0^q}{\lambda}\right)^{1/b} \tag{6.44}$$

This means that a certain minimum study effort must be maintained, as one's experience confirms. Notice how we have achieved some measure of validation of the model without as yet considering the numerical values of any of the parameters. Algebraic arguments such as this represent a powerful solution technique.

Note also that the solution has been obtained almost entirely manually with virtually no assistance from DERIVE. The reader may consider it strange to conclude a chapter in a book about modelling with DERIVE with an example where it can't offer much assistance. The authors believe that one should not lose sight of the fact that sometimes use of software (even DERIVE!) will require significant intelligent intervention on the part of the modeller.

This problem has also been considered by Cheung *et al.* [10] and a similar solution is presented in their paper.

Further Work

The reader is encouraged to develop a model based on the study strategy (b), namely, to study as hard as possible for as long as possible. Although you might think that this is an unreasonable approach, it does approximate the 'swot' situation!

The substitution represented by equation (6.39) can again be applied to equation (6.35) to give

$$\frac{1}{q} Z^{\frac{1}{q}-1} \frac{\mathrm{d}Z}{\mathrm{d}t} = \lambda q \, \mathrm{e}^{qmt} \left(K \mathrm{e}^{mt}\right)^{a} \mathrm{E}^{b}$$

whence

$$E(t) = \left(\frac{1}{\lambda q}\right)^{1/b} \left(\frac{\mathrm{d}Z}{dt}\right)^{1/b} \mathrm{e}^{-qmt/b} \tag{6.45}$$

where the study effort expended is now a function of time.

Remember that it is not $E(t)$ itself which is of prime interest but, rather, the conditions which will minimize the effort integral

$$\int_0^T E(t)\,dt$$

This type of problem falls within the mathematics of variational methods. The condition for the integral to take its minimum value is

$$\frac{\partial E}{\partial Z} = \frac{\mathrm{d}}{\mathrm{d}t}\left(\frac{\partial E}{\partial W}\right) \qquad \textit{where} \quad W = \frac{\mathrm{d}Z}{\mathrm{d}t}$$

Performing the analysis gives the optimal study effort (E_{opt}) for this model as

$$E_{\text{opt}} = \left(\frac{\delta}{\lambda w} \frac{\left(K_e^{q} \, \mathrm{e}^{qmT} - K_0^{q}\right)}{\left(\mathrm{e}^{qmT/w} - 1\right)}\right)^{1/b} \cdot \mathrm{e}^{qmt/w} \tag{6.46}$$

where δ is an arbitrary constant. The detail of the analysis is presented in [10].

Interpretation and Validation

With solutions now available for two different models of the study strategy it is appropriate to compare the results. Recall that for constant study effort

$$E_0 = \left(\frac{c\left(K_e^q \, e^{qmT} - K_0^q\right)}{\lambda\left(e^{qmT} - 1\right)} \right)^{1/b}$$

For unlimited study effort the corresponding result is given by equation (6.46).

How fortunate that there are many factors in common (who would have predicted that?) so that

$$\frac{E_{\text{opt}}}{E_0} = \left(\frac{e^{qmT} - 1}{w\left(e^{qmT/w} - 1\right)} \right)^{1/b} \cdot e^{qmt/w} \qquad (6.47)$$

Notice that this result is independent of λ, which means we have one less parameter to worry about when estimating parameter values. The ratio only depends on the poor memory factor (m), q and w (or, in terms of the parameters originally introduced, a and b which arose in the model we adopted for knowledge acquisition). What does the result mean for the student?

Provided that the individual can estimate his/her level of uniform study effort then equation (6.47) could be used to estimate his/her optimal study strategy. Its use would also require the student to conduct some trials in order to estimate the memory factor m and to research the educational psychology literature to ascertain reasonable values for a and b. The authors of the original paper used $a = 0.5$, $b = 0.5$ and $m = 0.25$. You could investigate the behaviour of (6.47) for other values of these parameters and for courses of different durations (T).

How might you modify the models to take account of any unfortunate overloadings of homeworks and courseworks from other courses?

As mentioned earlier, the unlimited effort model is unreasonable for all but the most enthusiastic students. A more reasonable further model would be one based on an upper limit to the study effort. The interested reader is again referred to the referenced paper for the development of this model.

References

[1] Edwards D. and Hamson M., *Guide to Mathematical Modelling*, Macmillan (1989).

[2] Burghes D.N. and Wood A.N., *Mathematical Models in the Social, M* *and Life Sciences*, Ellis Horwood, Chichester (1990).

[3] Dearden J.C. and Townend M.S., 'Mathematical modelling of the drug process – incorporation of realistic transfer rate constants and dissolution step', in *QSAR and Strategies in the Design of Bioactive Compounds*, edited by J.K. Seydel, VCH, Weinheim, Germany (1985).

[4] Townend M.S., *Mathematics in Sport*, Ellis Horwood, Chichester (1984).

[5] Hughes D.E., 'Some mathematics and physics of ball games', *School Science Review*, September (1985).

[6] Open University Course MST204 Mathematical Models and Methods, Unit 4.

[7] Mehta R.D., 'Aerodynamics of sports balls', *Ann. Rev. Fluid Mech.*, **17**, 151–89 (1985).

[8] Schwartzenbach J. and Gill K.F., *System Modelling and Control*, 3rd edition, Arnold (1992).

[9] Stroud K.A., *Further Engineering Mathematics*, 2nd edition, Macmillan (1994).

[10] Cheung B. K. S., Lee S. M. and Tang W. I., 'Student study problem revisited – A simple optimal control problem', *IMA. Bulletin*, **27**, 121–125 (1991).

7

Modelling With Recurrence Equations

7.1 INTRODUCTION

In the preceding chapter we investigated a variety of problems for which it was appropriate to develop continuous models, i.e. models based on the differential calculus. It must be emphasized that it is not appropriate to treat all situations as being continuous, and in this chapter we investigate some problems which require an alternative discrete model approach based on recurrence equations.

Recalling that the differential calculus is based upon infinitesimal changes in a variable, the decision as to which model is appropriate in any given situation is determined by deciding whether the variables involved can be regarded as being continuous or discrete. Furthermore, if they are considered to be discrete could they nevertheless be reasonably approximated by a continuous model? Whether or not you have made the right choice is determined by the validity of your solution. The decision process was illustrated in some detail in the introduction to the Indian population problem presented in Chapter 6.

Although today we are probably far more accustomed to models based on calculus rather than recurrence equations, it is interesting to note that recurrence equations significantly preceded the introduction of the calculus. One of the classic problems involving recurrence equations is due to Leonardo of Pisa (1175–1250), who was also known as Fibonacci, whereas Newton did not develop the calculus until the late seventeenth century. We shall use Fibonacci's problem to introduce the idea of a model involving recurrence equations.

The Problem

In 1202 Fibonacci published his book *Liber abbaci* in which he posed the following problem:

> How many pairs of rabbits will be produced each month, beginning with a single pair, if every month each 'productive' pair bears a new pair which becomes productive from the second month on?

Setting up a Model

In order to develop a solution to this problem we begin by numbering the months, 1,2,3,...,n and denote the number of pairs of rabbits alive at the beginning of the nth month by u_n. Following Fibonacci we shall also make the assumption that deaths do not occur. While this is obviously unrealistic we shall nevertheless follow Fibonacci's derivation because of its historical interest.

Formulate a Mathematical Model

The number of pairs alive at the beginning of month $n+1$ is then given by

$$u_{n+1} = \begin{array}{l}\textit{the number of pairs}\\ \textit{alive at the beginning}\\ \textit{of month } n\end{array} + \begin{array}{l}\textit{the number of}\\ \textit{pairs born during}\\ \textit{month } n\end{array}$$

The number of pairs born during month n comes from pairs over one month old, i.e. from the u_{n-1} pairs alive at the beginning of month $n-1$. Hence,

$$u_{n+1} = u_n + u_{n-1} , \qquad n = 2,3, \ldots \tag{7.1}$$

We can also add the information that $u_1 = 1$ (Fibonacci started with one pair) and $u_2 = 1$ (because rabbits are not considered productive during their first month of life).

Equation (7.1) is known as a recurrence equation. By using it recurrently with the starting values $u_1 = u_2 = 1$ the build up of the rabbit population can be investigated as n increases. For example, setting $n = 2$ in equation (7.1) gives

$$u_3 = u_2 + u_1 = 2$$

and setting $n = 3$ gives

$$u_4 = u_3 + u_2 = 3$$

Continuing in this recurrent manner the rabbit population in the first few months is given by the sequence

1, 1, 2, 3, 5, 8, 13, ...

which is known today as the Fibonacci sequence.

Solution of the Mathematical Model

The recurrence equation (7.1) is known as a second-order recurrence equation because the difference between the greatest and least suffices is two; recurrence equations are classified according to their order. It is also known as a linear relation because none of the terms in equation (7.1) is raised to a power greater than unity. DERIVE contains a utility file RECUREQN.MTH which can be used for solving recurrence equations. Section 9.9 of the utility files Help section lists the six types of recurrence equation whose solutions can be found using DERIVE, see Fig. 7.1.

```
                    Utility File Functions (Chapter 9)

9.9  RECUREQN.MTH - Solving Recurrence Equations
  LIN1_DIFFERENCE (p,q,x,x0,y0) := solves y(x+1)=p(x)y(x)+q(x), y(x0)=y0
  RECURRENCE1 (r,x,y,x0,y0,n) := n steps of y(x+1)=r(x,y(x)), y(x0)=y0
  GEOMETRIC1 (k,p,q,x,x0,y0) := solves y(k*x)=p(x)y(x)+q(x), y(x0)=y0
  CLAIRAUT_DIF (p,q,d,x,y,c) := solves p(y-xd)=q(d), d represents y(x+1)-y(x)
  LIN2_CCF (p,q,r,x,c1,c2) := solves y(x+2)+p*y(x+1)+q*y(x)=r(x)
  LIN2_CCF_BV (p,q,r,x,x0,y0,x2,y2) := solves y(x+2)+p*y(x+1)+q*y(x)=r(x)

9.10  APPROX.MTH - Pade Rational Approximation
  PADE(y,x,x0,n,d):=approx y(x) near x=x0, n=numr deg, d=denr deg, n=d or d-1

9.11  EXP_INT.MTH - Exponential, Log, Sine, & Cosine Integrals:  use approX
  euler_gamma := Euler's constant to 25 significant digits
  EI(x,m):=m terms of series for exponential integral INT(#e^-t/t,t,-x,inf), x>0
  LI(x,m):=m terms of series for logarithmic integral INT(1/LN t,t,0,x), x>1
  EN (n,z) := nth exponential integral INT(exp(-zt)/t^n,t,1,inf), RE(z)>0, n>=0
  EN_ASYMP (n,z,m) := m+1 terms of asymptotic series for EN(n,z), |z| large
  E1(z,m) := m terms of series for exponential integral E1(z)=EN(1,z)
  SI(z) := sine integral INT(SIN(t)/t,t,0,z)
  CI(z) := cosine integral INT(COS(t)/t,t,0,z), |phase z| < pi

HELP UTILITY: Next Previous Resume

Enter option
                                        Free:100%            Derive Algebra
```

Figure 7.1 DERIVE's help file for recurrence equations

Comparison of equation (7.1) with those shown in Fig. 7.1 reveals two major differences between the algebraic and DERIVE presentations of recurrence equations. Firstly, DERIVE does not support a suffix notation. Secondly, notice in Fig. 7.1 that there are no arguments which involve a subtraction. The connection between the equation (7.1) and the utility file notation is the equivalence

$$u_n \equiv y(x)$$

Equation (7.1) must therefore be re-written to avoid suffices such as $n-1$ and this is achieved simply by incrementing all the suffices by one to give

$$u_{n+2} = u_{n+1} + u_n$$

which can then be rearranged as

$$u_{n+2} - u_{n+1} - u_n = 0 \qquad n = 1,2, \ldots \tag{7.2}$$

and compared with the DERIVE format

$$y(x+2) + p(x)\cdot y(x+1) + q(x)\cdot y(x) = r(x) \tag{7.3}$$

The next question is to decide which of the second-order linear utility files LIN2_CCF and LIN2_CCF_BV is appropriate to the solution of equation (7.3).

Since we are told that $u_1 = 1$ and $u_2 = 1$, then we are in possession of boundary values (BV) so that it is LIN2_CCF_BV which is required. We can make the identification

$$p(x) = -1,\;\; q(x) = -1,\;\; r(x) = 0,\;\; x_0 = 1,\;\; y_0 = 1,\;\; x_2 = 2,\;\; y_2 = 1 \tag{7.4}$$

Solution of equation (7.1) or, equivalently, equation (7.2) can then be achieved with DERIVE by first loading the utility file RECUREQN.MTH and then Authoring and Simplifying expression 2 in Fig. 7.2 to produce the solution in the form represented by expression 4 of that same diagram.

1: "Generating the Fibonnaci sequence. Author and Simplify"

2: LIN2_CCF_BV(-1, -1, 0, x, 1, 1, 2, 1)

3: "to obtain"

4: $$2^{-x}\left[\frac{\sqrt{5}\,(\sqrt{5}+1)^x}{5} - \frac{\sqrt{5}\,(\sqrt{5}-1)^x\,\mathrm{COS}(\pi\, x)}{5}\right]$$

5: "To obtain the sequence Author and Simplify vector(#4,x,1,10,1)"

6: $$\mathrm{VECTOR}\left[2^{-x}\left[\frac{\sqrt{5}\,(\sqrt{5}+1)^x}{5} - \frac{\sqrt{5}\,(\sqrt{5}-1)^x\,\mathrm{COS}(\pi\, x)}{5}\right],\; x,\; 1,\; 10,\; 1\right]$$

7: [1, 1, 2, 3, 5, 8, 13, 21, 34, 55]

```
COMMAND: Author Build Calculus Declare Expand Factor Help Jump soLve Manage
         Options Plot Quit Remove Simplify Transfer moVe Window approX
Compute time: 1.9 seconds
Simp(6)                              Free:100%                Derive Algebra
```

Figure 7.2 Generation of the Fibonacci sequence using DERIVE

A table of results for $x = 1,2,3, \ldots$ (corresponding to the original month numbers $n = 2,3,4, \ldots$) is then obtained by Authoring and Simplifying the expression

VECTOR (#4, X, 1, 10, 1)

see expression 6 of Fig. 7.2, to produce the Fibonacci sequence shown as expression 7 in Fig. 7.2.

Further Work

The occurrence of the Fibonacci sequence is very widespread throughout nature. The family history of a male bee (called a drone) provides an interesting illustration.

Before attempting to model the spread of a drone's family tree, some background information may be useful. A drone has a mother but no father. Within the colony the queen bee lays both fertilized and unfertilized eggs; the former hatch into females (workers or queens) while the latter hatch to produce drones.

Starting with a drone he would have had a mother but no father, and this mother would have had both a mother and a father. These bees would have had...(and so on).

(i) Continue this family tree backwards for a further four generations.
(ii) Count the number of bees in each generation.
(iii) Write down the recurrence equation linking the number of bees in successive generations.

Your investigation should produce the Fibonacci sequence for part (ii) and the second-order recurrence equation (7.1) for part (iii).

The general solution of equation (7.1) turns out to contain a very interesting number, as we investigate next. Within the utility file RECUREQN.MTH the function LIN2_CCF will yield the general solution of equation (7.1) in terms of two arbitrary constants $C1$ and $C2$. With the Option Precision set to Exact, obtain the general solution of equation (7.1). The solution is seen to contain, *inter alia*, the terms $(1 \pm \sqrt{5})/2$ which are the solutions of the auxiliary equation

$$\lambda^2 - \lambda - 1 = 0$$

associated with the recurrence equation (7.1).

The quantity $(1 + \sqrt{5})/2$ is of particular interest and is known as the golden ratio. The golden ratio may be defined as the result of dividing a line so that the ratio of the whole is to one part as that part is to the other.

Consider the Fibonacci sequence again:

$$1, 1, 2, 3, 5, 8, 13, \ldots$$

and investigate the limit of the ratio of successive terms of the sequence, i.e.

$$1/1\,, \quad 2/1\,, \quad 3/2\,, \quad 5/3\,, \quad 8/5\,, \ldots$$

and compare your results with $(1 + \sqrt{5})/2$.

You can do this using DERIVE as follows. First Transfer, Load the utility file MISC.MTH which, *inter alia*, contains the function FIBONACCI(n) which Simplifies to the nth Fibonacci number. Then Author and approXimate the vector

VECTOR(FIBONACCI(n+1)/FIBONACCI(n), n, 1, 19)

which gives a vector of the values of the ratio for the first twenty Fibonacci numbers.

Both the Fibonacci sequence and the golden ratio can be identified in many diverse areas as the following brief summary indicates:

Fibonacci sequence – the number of petals on a flower head,

– the arrangement of leaves on stems of plants.

Golden ratio – the proportions of the human body (for example, the ratio of foot to navel distance to height to hairline),

– the geometry of the construction of the Egyptian pyramids and the small Aten Temple at Amarna,

– cycles of change in the stock market.

The interested reader is referred to [1] for further information.

7.2 SEAL CULLING

(Adapted from an exercise from the Open University course MST204 Mathematical Models and Methods.)

Introduction

This model is about the emotive environmental issue of seal culling. Seal culling is carried out periodically because man wishes to reduce the seal population in order to lessen the competition for the fish stocks on which he depends or which he exploits commercially.

The Problem

A model is required which will predict the change in the number of female seals within a colony over time in order to determine whether culling is necessary.

This is the sort of problem which probably requires the modeller to liaise with an expert at an early stage in order to establish data about the reproductive pattern of seals, death rates, typical colony size, and so on. To assist you in the development of a model the following information about the reproductive pattern and life cycle of seals is presented:

- Each productive female seal alive at the start of a year produces, on average, 0.75 new female seals in that year. (This is not a nonsense value, it merely means that on average they produce three female offspring every four years.)
- Females become productive in their second year and are able to produce new seals each year thereafter.
- The number of females who die in any one year is proportional to the number of females alive at the start of that year.
- The female count in one specific colony in two successive years was 100 and 120 respectively.

Apart from its use in providing initial values, the last piece of information above is also useful as it indicates that a continuous, calculus based model is inappropriate (the time step of one year is quite large compared with the lifetime of a seal).

Consequently, a discrete model is required and we thus seek to establish a recurrence equation to describe the changes in the numbers of females.

Setting up a Model

In order to assist the development of this recurrence equation some simplifying assumptions are needed. The effects of immigration to and emigration from the colony are therefore ignored, as are any killings by predators. We agree that these assumptions are not realistic but they keep the initial model simple and could be taken into account in later revisions.

Formulate a Mathematical Model

If u_k denotes the number of female seals alive at the start of year k show that the above assumptions, together with the lifecycle data supplied earlier, lead to the recurrence equation

$$u_{k+1} = (1-d)u_k + 0.75\, u_{k-1} \tag{7.5}$$

where d is a constant of proportionality, $0 \le d \le 1$, and is a measure of the death rate (deaths in year $k = du_k$).

Express equation (7.5) in a form suitable for solution using DERIVE.

Solution of the Mathematical Model and Interpretation

Investigate the DERIVE solution of equation (7.5) for different values of d. Your investigations should reveal that the female population only goes into a decline for large values of d, i.e. for very high death rates such as those consequent upon severe storms which disperse the seals' food stocks dramatically. For smaller values of d, corresponding to the more commonly occurring lower death rates, the female seal population increases steadily. Hence the need for man to cull the population and preserve the fish stocks for his own use.

Further Work

Investigate the behaviour of seal colonies to establish whether there is significant immigration/emigration to/from any specific colony. If necessary, suggest appropriate modifications to the model developed so far. What effect do they have on your solution? What about modifications to account for the effect of predators?

7.3 MORTGAGE REPAYMENTS

Introduction

Mortgage repayment provides a natural example of the occurrence of recurrence relations in our everyday lives. Consider, for example, the case of a family which has a £20 000 mortgage to be repaid over twenty-five years at 9% per annum.

First let us determine the monthly repayment needed to pay off the mortgage by the end of the twenty-five-year period. The fact that the debt is being paid off with a discrete amount of money at a regular time interval rather than on a continuous basis suggests that a recurrence equation approach is appropriate.

The 'mechanics' of a mortgage are as follows – the amount to be repaid has to cover both the outstanding loan and the interest which has to be paid on the loan; the total sum to be repaid is then divided into equal monthly repayments for the period of the loan.

Setting up and Formulation of a Mathematical Model

Suppose that the monthly repayment is denoted by £M and that the amount owed at the start of year r is £u_r ($r = 0,1,2, \ldots$); then

	Amount owed at start of year r+1	=	*Debt at start of year r*	+	*Year r interest on debt*	−	*Twelve monthly repayments*
i.e.	u_{r+1}	=	u_r	+	$0.09\,u_r$	−	$12M$

which can be rearranged as

$$u_{r+1} = 1.09\, u_r - 12M \tag{7.6}$$

together with the initial condition $u_0 = 20\,000$. Equation (7.6) is to be solved to find the value of M such that $u_{25} = 0$ (i.e. the mortgage has been paid off).

Solution of the Mathematical Model

Comparison of the first-order recurrence equation (7.6) with the functions offered in RECUREQN.MTH shows that LIN1_DIFFERENCE is appropriate. Referring to Fig. 7.3, expression 4 is the result of Authoring and Simplifying expression 2. Expressions 5 and 7 show the result of Managing and Substituting $x = 25$, approXimating the result and then soLving for M to give, after approXimation, the required repayment as £169.68 per month (a financial institution such as a building society certainly wouldn't round the repayment down!).

You could also tackle this problem directly using the Financial Functions provided within DERIVE.

```
1:   "Mortgage repayment models"

2:   LIN1_DIFFERENCE(1.09, - 12 m, x, 0, 20000)

3:   "Simplify then Manage and Substitute x=25"

      x LN(109) + 2 (1 - x) LN(5) + 2 (2 - x) LN(2)
     ê                                              (150 - m)     400 m
4:   ──────────────────────────────────────────────────────── + ─────
                               3                                   3

      25 LN(109) + 2 (1 - 25) LN(5) + 2 (2 - 25) LN(2)
     ê                                                 (150 - m)     400 m
5:   ─────────────────────────────────────────────────────────── + ─────
                               3                                      3

6:   "approXimate and soLve for monthly repayment m."

7:   m = 169.677
─────────────────────────────────────────────────────────────────────────
COMMAND: Author Build Calculus Declare Expand Factor Help Jump soLve Manage
         Options Plot Quit Remove Simplify Transfer moVe Window approX
Compute time: 0.2 seconds
Solve(5)                          Free:100%                 Derive Algebra
```

Figure 7.3 Mortgage repayment calculation using DERIVE

Further Work

For convenience in their home accounting, one of the authors rounds his mortgage repayment up to the nearest ten pounds (i.e. £170), while the other rounds up to the nearest integer multiple of twenty-five (i.e. £175). How much sooner would the mortgage be paid off under these two repayment plans? How sensitive is the mortgage repayment to changes in the interest rate ?

7.4 BUILDING A NEST-EGG – AND DISMANTLING IT!

The Problem

Many parents make some financial provision for their children from birth. A typical plan might be to invest £100 at birth at a constant rate of compound interest of 8%, with the condition that the account is not to be touched until the child's twenty-first birthday. Use DERIVE to determine the final value of the account (£503.38).

One parent realizes that over such a long period of time the value of the savings will be seriously eroded by inflation and decides to counter this by adding £20 to the investment every birthday. We next investigate the additional benefit of this approach.

Setting up a Model

If £u_0 denotes the value of the investment at birth, £u_1 the value at the first birthday, and so on, show that

$$\begin{aligned} u_0 &= 100 \\ u_1 &= 1.08 \times 100 + 20 \\ u_2 &= 1.08 \times u_1 + 20 \\ &= 1.08^2 \times 100 + 20\,(1.08 + 1) \end{aligned}$$

and so on.

Solution of the Mathematical Problem and Interpretation

Develop the expression for £u_{21}, the final value of the investment, and simplify the result using DERIVE's Summation facility (£1511.83).

Another person's strategy for reducing the long term effects of inflation is to add £5 to the investment on the first birthday, £10 on the second birthday, and so on, increasing the amount added by £5 every birthday. Use DERIVE to investigate the value of this scheme.

Further Work

Yet another scheme encountered by the authors is to deposit £K initially and then double the deposit each year thereafter. Use DERIVE to examine how this compares with the initial scheme discussed (deposit £100 and leave it to accrue interest for the twenty-one years). Assuming $K < 100$, for what value of K would the two schemes produce the same overall sum? With the 'doubling scheme' is it likely that in the later years the deposits required would be so large as to be prohibitive?

Of course, not everyone has the patience to wait for such a long time before their investment matures, nor indeed might it be appropriate. A more common scenario is the withdrawal of a fixed sum each year for the lifetime of the investment. This is the situation which might be encountered in paying school fees from the proceeds of the initial investment of a lump sum. The following exercise is typical.

School Fees Payment Plan

A young girl's parents inherit £20 000 and decide to invest it at a rate of 8% per annum in order to fund their daughter's education, for which the annual fees are £2500, to be paid at the end of each year.

(i) Show that this leads to the first-order recurrence equation

$$u_{n+1} = 1.08\, u_n - 2500 \qquad n = 0,1,2,\, \dots$$

where £u_n is the value of the investment at the start of year n, and $u_0 = 20\,000$.

(ii) Does the money run out? If so, when?

The answer to part (ii) enables the parents to decide whether or not they need to make any further financial provision for the later years of their daughter's education.

(iii) Of course, school fees, like anything else, are affected by inflation and are thus unlikely to remain at £2500 per annum for the duration of the child's education. Suppose that the school decides to increase its fees by 4% per annum to compensate for inflation. What happens to the parents' plans now?

7.5 WORMS

(Source: *Mathematics and Statistics for Bioscientists*, Eason G., Coles C.W. and Gettinby G., Ellis Horwood (1980)).

When births and deaths are allowed for in population modelling, be it a continuous or a discrete model, one of several things can happen to the population. It may

(i) increase without limit or, more realistically, increase to some finite limiting value,
(ii) go into decline, leading ultimately to extinction,
(iii) more unusually, oscillate between an upper and a lower bound.

These various outcomes are illustrated by the following exercise about worms.

Consider a population of worms living under 'ideal' conditions in a laboratory (with sufficient food, no competition for space, no immigration ...). Assuming that their life cycle takes three years, then the total number u_k present at any time k will depend on the number of viable eggs laid two years ago and the number surviving the larval stage, which has a duration of one year. This can be modelled by the linear second-order recurrence equation

$$u_{k+2} = a\, u_{k+1} + b\, u_k$$

where a and b are both positive constants (b is related to the dependence on the number of viable eggs and a to the dependence on the number surviving the larval stage).

Find the solution for the following three cases:

(i) $a = 0.5\,, b = 1.5$
(ii) $a = 0.5\,, b = 0.1875$
(iii) $a = 0\,, \quad b = 1$

assuming that in each case $u_0 = 8$ and $u_1 = 10$.

Use the Calculus, Limit facilities to investigate the behaviour of your solutions in each case after a long period of time has elapsed. In case (iii) a failure is encountered because of the oscillatory nature of the solution $(9 - \cos(\pi x))$. All is not lost; remember that the cosine function oscillates between the values +1 and −1 ... .

Your investigation should show that in case (i) the population becomes infinite, in case (ii) it declines to zero and in case (iii) it oscillates between the values eight and ten.

7.6 DISH WASHING

Introduction

In the domestic situation of dishwashing, the pile of dirty dishes is usually small in number, even though it may not be considered so by the human dishwasher! Consequently, each article that is washed represents a significant element of the total number of dirty articles and hence it is inappropriate to attempt to develop models of domestic dishwashing which are based on an infinitesimal, calculus approach. Instead, we shall investigate the development of a model based on recurrence equations.

The Problem

First, consider the washing-up process itself. The water in the washing-up bowl must be hot enough to clean the dirty dishes but cool enough to put your hands in (thinks – how would the wearing of rubber gloves affect things?). During the washing-up period the water temperature reduces until eventually it is too cool to be an effective cleanser.

The authors' considerable experience of washing-up has established that there are at least two washing-up protocols as follows:

(a) Immerse each article separately, wash it and remove it before inserting the next article.
(b) Put all the dirty dishes in the water together and remove each immediately after it has been washed.

We shall outline the development of a model based on protocol (a).

The purpose of this case study is to develop a model which can be used to predict the number of dirty dishes which can be washed using one bowl of washing-up water.

Setting up a Model

The physical processes which take place in the washing-up cycle can be identified as follows:

(i) Each item (at ambient temperature) is immersed in the water and becomes hotter:

conduction and convection.

(ii) The item is then washed, any grease being melted off it:

conduction to the item.

(iii) The item is removed from the bowl. The water in the bowl has cooled:

convection and radiation from the water surface.

This is quite a complicated procedure and some simplifying assumptions would now be helpful.

Suppose we make the assumption that on immersion into the water the temperature of the dirty item instantly changes from ambient to the current water temperature. This means that we do not need to consider process (i) above. (Process (i) can be accounted for; its inclusion would require values for the thermal conductivity of the crockery and the heat transfer coefficient for convection in water.)

Since heat is a form of energy it seems reasonable to base our model on the principle of conservation of energy, in this case heat energy.

Formulate a Mathematical Model

The model can be expressed as

			heat energy of item on removal	+	*heat energy of water after item removed*	
Heat energy of item before immersion	+ *heat energy of water before item inserted*	=				
		+	*energy losses by convection*	+	*energy losses by radiation*	(7.7)

Before defining our list of variables and considering any further assumptions some technical comment may be necessary about the individual terms in the above equation. The following are well known:

- The heat energy contained within a body of mass m at temperature T (degrees Kelvin) is mcT, where c is known as the specific heat capacity and is a property of the material of the body.
- The rate of energy loss by convection is given by

$$E_c = h\,A\,(T - T_a)$$

  where h is the heat transfer coefficient, A is the surface area, T(K) is the body temperature and T_a(K) is the ambient temperature.
- The rate of energy loss by radiation is given by

$$E_r = \varepsilon\,\sigma\,A\,(T^4 - T_a^4)$$

  where σ is the Stefan Boltzmann constant and ε measures the emissivity of a surface (for water $\varepsilon = 0.95$ approximately and $\sigma = 5.67 \times 10^{-8}$ Wm^{-2} K^{-4}).

In order to develop equation (7.7) into a quantitative model some modelling assumptions are needed. We shall assume the following:

- The amount of water in the bowl is constant (not quite true as there will be some splashing and drips off the dishes. Also there may be a small evaporation loss).
- All dishes, etc., are initially at the ambient temperature.
- All dishes, etc., are of the same mass.
- All dishes, etc., are immersed for the same time, Δt (seconds) (which is also long enough for the plate to reach the same temperature as the water).

As the final step before the development of the recurrence equation based model we must define all the variables (and their units):

average mass of an item of crockery	m_p	(kg)
mass of water	m_w	(kg)
specific heat capacity of plate	c_p	($Jm^{-3}K^{-1}$)
specific heat capacity of water	c_w	($Jm^{-3}K^{-1}$)
ambient temperature	T_a	(K)
water temperature	T_n	(K)
initial water temperature	T_0	(K)
final effective water temperature	T_f	(K)
water surface area	A	(m^2)
heat transfer coefficient from air to water	h	($Wm^{-2}K^{-1}$)
number of dishes washed	n	

In terms of these variables, show that equation (7.7) can then be expressed as

$$(m_p c_p + m_w c_w)T_{n+1} = (m_p c_p + hA\Delta t)T_a + (m_w c_w - hA\Delta t)T_n - \varepsilon\sigma A(T_n^4 - T_a^4)\Delta t \quad (7.8)$$

Solution of the Mathematical Model

The presence of the term T_n^4 means that the recurrence equation is highly non-linear and is certainly not soluble using functions provided in the DERIVE utility file RECUREQN.MTH.

Following standard modelling practice (namely, can we simplify the model and yet still retain its plausibility?) we investigate the possibility of neglecting the non-linear term. We would be able to do this if we could establish that the magnitude of this radiation term was negligible compared with the convection term in equation (7.8).

A simple experiment will establish the maximum water temperature which your hands can tolerate (T_0 K). Using your value of T_0 as the water temperature and assuming that $h = 100\ Wm^{-2}K^{-1}$, determine the value of the ratio E_r/E_c. Your results should suggest that E_r is negligible compared with E_c, which gives reasonable grounds for ignoring the radiation term in equation (7.8), resulting in

$$(m_p c_p + m_w c_w)T_{n+1} = (m_w c_w - hA\Delta t)T_n + (m_p c_p + hA\Delta t)T_a \quad (7.9)$$

In order to pursue this, some data values are now required. You will need to measure the cross sectional area of your washing-up bowl (A m^2), the mass of the washing-up water in the bowl (m_w kg), the average mass per crockery item (m_p kg), the ambient room temperature (T_a K) and the average immersion time (Δt s). Finally, continue washing the dishes until you establish the value of the minimum effective water temperature (T_f K). (All the temperatures were originally measured in degrees Kelvin because that is the unit required in the radiation term. However, now that the radiation term has been neglected the temperatures could equally well be measured in degrees Celsius.) Values of c_p, c_w and h are available from the technical literature (see, for example, [2]) but are quoted here for convenience:

$$c_p = 600 \text{ Jm}^{-3}\text{ K}^{-1}$$

$$c_w = 4200 \text{ Jm}^{-3}\text{ K}^{-1}$$

$$h = 100 \text{ Wm}^{-2}\text{ K}^{-1}.$$

With all your parameter values inserted, equation (7.9) can then be arranged into the form

$$T_{n+1} = \alpha\, T_n + \beta \tag{7.10}$$

together with the initial condition that $T = T_0$ when $n = 0$.

Next, you will need to decide which of the functions available in RECUREQN.MTH is appropriate to the solution of the problem posed by equation (7.10). The available functions suggest the following two possible approaches to produce the solution.

(a) LIN1_DIFFERENCE
This will provide the solution to (7.10). Set T_f equal to this solution and then solve for n, the number of items of crockery which can be effectively washed.

(b) RECURRENCE1
Guess a value for n and then use this function to produce a table of temperature values. Refine your estimate of n as necessary until your DERIVE output reaches the temperature T_f.

Do your results agree with your washing-up 'experience'?

Further Work

How is the value of n affected if a larger volume of washing up water is used? How are the results affected if washing protocol (b) is adopted?

7.7 TREASURY MODELS

In order that a country will not go bankrupt, the officers in its Treasury will have developed sophisticated models to predict the dependence of the national income on such things as

(i) government expenditure on defence, education, health, and so on,
(ii) private investment in industry, commerce, etc.,
(iii) consumer expenditure on goods and services,
(iv) imports and exports of goods and services,

and you can probably add further factors to the above list.

The Problem

The sort of information which the Treasury officials require from their model is whether the national income grows, decays or oscillates over time. How is it affected by the level of government expenditure? Could government expenditure be used to control the behaviour of the economy?

Setting up a Model

One of our first decisions in developing such a model is to decide whether it should be discrete or continuous. While it is true that the amount of money in the government coffers (or your own pocket) changes continuously from day to day, the time unit often used by financial planners is one year (we are all familiar with *annual* subscriptions, the *yearly* form from The Inland Revenue, etc.), and so it is proposed to develop a discrete model, which will of course lead to a recurrence equation. If you disagree with this assumption (one of the privileges of being a mathematical modeller!), then your continuous model will result in a differential equation.

We begin by introducing the notation:

I_k, the national income
S_k, government expenditure on defence, etc.
P_k, the amount of private investment
E_k, consumer expenditure
T_k, net income from imports and exports

each being measured in the same, appropriate 'mega' unit of currency. The suffix k denotes the kth time period (the unit of time being one year).

Formulate a Mathematical Model

The model outlined in the opening paragraph can then be expressed in the form

$$I_k = E_k + P_k + S_k + T_k$$

For simplicity we shall assume that the net foreign trade is zero (i.e. value of exports = value of imports each year) so that $T_k = 0$ for all k and our model becomes

$$I_k = E_k + P_k + S_k \tag{7.11}$$

We now need some submodels for consumer expenditure and private investment as, obviously, these are affected by the state of the national income. A buoyant economy, always changing for the better, will stimulate both purchasers and investors.

First consider consumer expenditure. We shall assume that it is directly proportional to the value of the national income in the previous year as purchasers will probably want to see that the economy is performing well before they commit themselves to buying, i.e.

$$E_k = a\, I_{k-1} \quad , \ a \text{ is a positive constant} \tag{7.12}$$

As far as private investment is concerned, investors are attracted not so much by high consumer spending but by a big change in the amount of consumer spending from one year to the next (why?) so a reasonable model would be

$$P_k = b\,(E_k - E_{k-1}) \quad , \ b \text{ is a positive constant} \tag{7.13}$$

Finally, if we assume for simplicity that government expenditure is constant from one year to the next (so that $S_k = S$ (constant) for all k), show that equations (7.11)–(7.13) can be combined to produce the following linear second-order, non-homogeneous recurrence equation:

$$I_{k+2} - a(1+b)\, I_{k+1} + ab\, I_k = S \tag{7.14}$$

the suffices having each been increased by two units to bring them into line with the notation used by DERIVE. Obviously, the solution of (7.14) will depend on the values of a, b, S and the two initial values of the national income which we may need to provide (why two?).

Although it is tempting to substitute values for a, b and S into (7.14) immediately and then attempt solution via DERIVE, it is usually possible to obtain much more insight into one's problem by attempting to find the general solution and then investigating its behaviour.

Solution of the Mathematical Model

When you study the contents of the utility file RECUREQN.MTH you should conclude that the appropriate function to use to solve (7.14) is LIN2_CCF. Author the appropriate expression and then Simplify it to produce the extremely unwieldy general solution shown as expression 4 in Fig. 7.4. (We hope you remembered to Load the utility file first!)

```
1:   "modelling the economy"

2:   "General solution: Simplify the expression"

3:   LIN2_CCF(- a (1 + b), a b, g, x, c1, c2)

          (x - 2) / 2                                  2
4:   (a b)            [COS(x ATAN(√(a (4 b - a (b  + 2 b + 1))))) [g Σ ―――――
                                                                   x
                                                                     SIN((x +

5:   "For solution of assoc.homog.eqn. Simplify"

6:   LIN2_CCF(- a (1 + b), a b, 0, x, c1, c2)

       x/2                                2
7:   (a b)    (c2 COS(x ATAN(√(a (4 b - a (b  + 2 b + 1)))) + c1 SIN(x ATAN(√(a

COMMAND: Author Build Calculus Declare Expand Factor Help Jump soLve Manage
         Options Plot Quit Remove Simplify Transfer moVe Window approX
Compute time: 0.4 seconds
Simp(6)                              Free:90%                    Derive Algebra
```

Figure 7.4 DERIVE's analytical solution of the treasury modelling problem

What can we do now?

Without going into the detail of the solution of recurrence equations the solution of (7.14) is made up of a particular solution (given by $S/(1-a)$) and the solution of the associated homogeneous equation

$$I_{k+2} - a(1+b)\,I_{k+1} + ab\,I_k = 0 \quad . \tag{7.15}$$

Prompted by the fact that the right-hand side of equation (7.14) is a constant, the particular solution was found by adopting a trial solution of a similar form i.e. $I_k = C$, a constant, to produce the result $C = S/(1-a)$.

The solution of equation (7.15) can be found using DERIVE and is shown as the much more manageable expression 7 in Fig. 7.4.

Now we are ready to investigate the behaviour of the national income (I_k) as time progresses (i.e. as k increases), for different values of a and b. Notice, too, that with this approach there is no need to use the recurrence equation plus some initial values (for I_0 and I_1) to produce tables of numerical values of I_k. It is not always easy to infer the correct behaviour of a function just by looking at a table of values – the algebraic approach adopted here is potentially much more informative.

Interpretation

Provided that you avoid the value $a = 1$ (why? – hint, look at the particular solution), you can investigate the behaviour of the solution represented by expression 7 (Fig. 7.4) as x increases (corresponding to increasing the time counter k) for any other values of a and b.

Is it reasonable to ignore the case $a = 1$? In the context of this problem the answer is 'yes' as no country would allow its consumer expenditure in one year to equal the total national income of the previous year (only a proportion less than 1 would be allowed).

Before actually substituting any specific values for a and b let's see if we can make any deductions about the behaviour of the solution from its algebraic form. By considering expression 7 (Fig. 7.4), which represents the solution of the associated homogeneous equation, show that one of its components is a bounded trigonometric function provided that

$$a\,(4b - a(b^2 + 2b + 1)) > 0 \tag{7.16}$$

and deduce that in this case the long term behaviour of the economy is governed by

$$\lim_{x \to \infty} (ab)^{x/2}$$

and hence is stable provided that

$$ab < 1 \tag{7.17}$$

By combining the inequalities (7.16) and (7.17) deduce that this model predicts a stable economy provided that

$$0.25\, a^2\, (1 + b)^2 < ab < 1 \tag{7.18}$$

You could confirm these results by using the Manage, Substitute commands to insert particular values of a and b into expression 7 and then use the Calculus, Limit facilities to investigate the behaviour of the national income as $x \to \infty$ (don't forget the particular solution).

For example, try

(i) $a = 0.5$, $b = 0.8$

(ii) $a = 0.6$, $b = 0.3$

(iii) $a = 0.8$, $b = 1.6$

Can you explain the result for the last case?

In both cases (i) and (ii), the solution tends to $S/(1-a)$, i.e. the national income is predicted to be stable and equal to some multiple of the government expenditure.

Further Work

One criticism (at least!) of this model can be levelled at the assumption of constant government expenditure. A government may well decide that if, say, private investment is high in some sectors of the economy, then it can afford to reduce its own expenditure in those sectors. We might model this as

$$S_k = \lambda / P_k$$

where λ is a positive constant. What effect does this have on the original model represented by equation (7.11)? You should find that the replacement for equation (7.11) is non-linear and so cannot be solved using any of the methods contained in RECUREQN.MTH.

Nevertheless, it is possible to develop a numerical solution once you have specified a set of values for a, b and λ. Then, provided you have two initial values to start the recurrence relation (say $I_0 = 2$ and $I_1 = 2.5$ to represent an economy which is initially buoyant), you can use the recurrence relation to predict I_2, I_3, I_4, ... Investigate!

7.8 SATELLITE TELEVISION

The Problem

A company which provides a satellite television installation service is considering whether or not to open a branch in a small town consisting of about 5000 dwellings.

In order to establish whether the project would be commercially successful the company wishes to develop a model which will predict the demand for satellite television in the town. In particular it would like to be able to estimate both the number of systems likely to be installed per year and the associated installation cost. Develop a model for use by the company.

Some basic business concepts may be useful in helping you to get started:

(i) The demand for a commodity goes down as the price goes up.

(ii) If sales go up then the workload on the company's engineers will be increased. This can be managed by offering the engineers overtime or by appointing extra staff or, as we propose here, this pressure can be relieved by making any price increase proportional to the increase in demand.

(Incidently, the overtime versus extra employees' situation makes an interesting modelling problem in its own right.)

Setting up a Model

The company's experience of operating similar branches elsewhere is that the demand for its system is linearly related to the installation price by the equation

$$Demand \;= 5000 - 4 \times installation\ price$$

We shall also assume that any price increase in a specific year is proportional to the increase in demand experienced in the previous year.

Already, we see the need to introduce a more mathematical notation. Suppose we let

p_r = *price of installation in year r* (pounds)

d_r = *number of installations in year r*

Formulate a Mathematical Model

For a given year r, show that the assumption that the proposed price increase in year $r + 1$ is proportional to the increase in demand experienced in year r leads to the equation

$$p_{r+1} - p_r = \lambda\,(d_r - d_{r-1}) \tag{7.19}$$

where λ is the constant of proportionality, and that the demand–installation price relationship can be expressed as

$$d_r = 5000 - 4\,p_r \tag{7.20}$$

By combining equations (7.19) and (7.20) show that the installation price satisfies the second-order recurrence equation

$$p_{r+1} + (4\lambda - 1)p_r - 4\lambda\,p_{r-1} = 0 \tag{7.21}$$

Solution of the Mathematical Model

The company expects to set the installation cost at £200 initially and to raise this to £220 after the first year. Investigate the commercial success or otherwise of the venture for different values of λ – why is it appropriate to only use fairly small values of λ (suggested values are $\lambda = 0.05, 0.10, 0.15$)?

Interpretation

Your results should show that the number of sales stabilizes in each case. In fact the sales stabilize quite quickly (by year 6) for $\lambda = 0.05$ but take significantly longer for the higher values of λ. In these latter cases we would certainly have to consider the effects of inflation – indeed, this simple model would probably have become redundant due to changes in satellite television technology or leisure patterns. Notice, too, that for the case $\lambda = 0.05$ the proposed first year price increase to £220 has been reduced for the following year.

The model would prove useful to the satellite television company in predicting what might happen if a sudden price increase were implemented (perhaps in an attempt to ease a cash flow problem) or if a change were made in the policy for determining the annual price increase (equation (7.19)).

Critical appraisal of a solution is an important part of the modelling process. It should be noted that the recurrence equation (7.20) implies an upper limit on the price of installation. Do you think this is reasonable?

7.9 THE GROWTH OF THE POPULARITY OF DERIVE

Soft Warehouse (the producers of DERIVE), Prentice Hall (the publishers of this text and other DERIVE-related texts) and of course we as authors, all hope to see the growth in popularity and use of DERIVE as a mathematical assistant in education and industry throughout the world. Here we attempt to model such growth.

Setting up a Model

Suppose that x_n represents the number of regular users of DERIVE after n equal intervals of time (say, years). Intuitively, it would seem likely that the more people using DERIVE, the more they will spread the good news to others so that they, too, will become regular users. This would suggest a recurrence equation of the form

$$x_{n+1} = x_n + f(x_n) \qquad (7.22)$$

where $f(x_n)$ is some function of x_n to be specified. Naturally, we would hope that $f(x_n) \geq 0$ for all n (why?) and we are assuming that 'once a DERIVE user, always a DERIVE user'.

Formulate a Mathematical Model

Before specifying a form for $f(x_n)$, let us consider the market for DERIVE. This could be the entire world population from school age upwards, which we shall denote by P. Thus x_n / P represents the fraction of the potential market for DERIVE in a given year. It will be convenient to re-scale by choosing $P = 1$ and re-define x_n as the fraction of the market for DERIVE after n years.

An initial thought for $f(x_n)$ might well be $f(x_n) = kx_n$ for some positive constant k. Hence (7.22) becomes

$$x_{n+1} = (1+k)x_n \tag{7.23}$$

Solution of the Mathematical Model and Interpretation

Use DERIVE to examine the sequence x_1, x_2, ... generated by this recurrence equation for various values of k, say $k = 0.2$ and 0.4, starting with an initial value $x_0 = 0.05$ (why not $x_0 = 0$?).

This can be conveniently done by Authoring the statements

1. ans(n) : = iterate ((1+k)x, x, 0.05, n)
2. vector ([n, ans(n)], n, 1, 20)

and then approXimating expression 2.

Ans(n) contains the value of x_n and expression 2 produces a vector of the results [n, x_n] for values of n from 1 to 20. Use the Manage, Substitute commands to insert values for k in expression 1. The approXimate form of expression 2 can then be plotted using an appropriate scale (say $x = 3$, $y = 0.4$) and Centering the graph as usual. A typical DERIVE plot for $k = 0.2$ and $k = 0.4$ with n values along the horizontal axis and x_n values along the vertical axis is shown in Fig. (7.5), in which the lower curve corresponds to $k = 0.2$. (Note that the points (n,x_n) have been joined by lines using the Options, State command and selecting Connected under Mode.)

A glance at the plots obtained should soon convince you that something is seriously wrong with this model. It predicts unlimited growth for x_n as n increases and yet x_n is limited to values less than or equal to unity (remember the original scaling).

Readers familiar with ordinary differential equations should recognize that the continuous equivalent of (7.23) is

$$\frac{dx}{dt} = kx \ , \quad \text{where } k \text{ is termed the growth-rate}$$

For positive k this has a solution of the form $x = Ae^{kt}$, which is unbounded as $t \to \infty$. Hence the behaviour of the solution of the recurrence equation was not entirely unexpected.

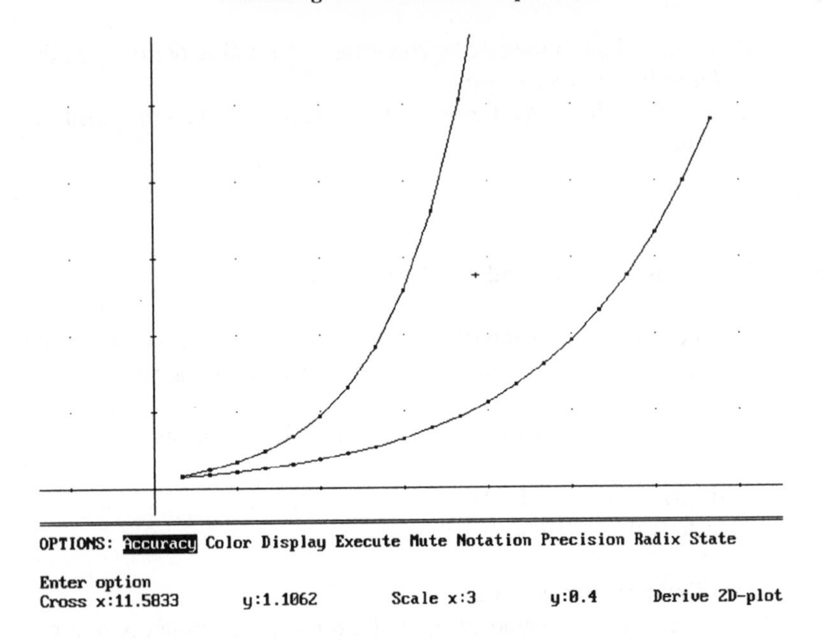

Figure 7.5 Unbounded demand curves for DERIVE

To produce a model that predicts a limit to growth, then we need a growth-rate which is not a constant, but a function of x_n that tends to zero as x_n tends to 1 (in this scaled case). Following the usual adage of 'keep it simple', we might now try a linear form for the growth-rate and propose a growth-rate of the form $k(1-x_n)$ leading to the recurrence equation

$$x_{n+1} = x_n + k(1-x_n)x_n \tag{7.24}$$

i.e.

$$x_{n+1} = (1+k)x_n - kx_n^2 \tag{7.25}$$

Use DERIVE expressions similar to those above for constant growth-rate to generate a sequence of values $[n, x_n]$ for x_n given by (7.25). Use $k = 0.2$ and $k = 0.4$ as before. Plot the sequence and observe that this time the limiting value of 1 is approached as anticipated. Generate about forty sequence values and scale the plot accordingly. Observe that the $k = 0.4$ plot approaches the limiting value more quickly.

For $k = 0.2$, the model predicts that the market becomes saturated after approximately thirty-six years, whereas for $k = 0.4$, this happens after approximately eighteen years. Is this factor of ½ for a doubling of k just a coincidence? Try plotting $k = 0.1$, $k = 0.8$ to see if these plots confirm/refute your conjecture.

Comment/Further Work

Equation (7.24) is known as the logistic equation and has been developed and used since the middle of the nineteenth century to model population growth. Having plotted the sequence for $k = 0.1, 0.2, 0.4, 0.8$, you may be forgiven for thinking (wrongly) that the sequence is fairly predictable. Try generating and plotting the sequence for $k = 1, 2, 3, 4, \ldots$. Clearly, for $k > 1$, values of $x_n > 1$ are obtained which invalidate the model. None the less, the plots are interesting in that they show that the solutions are chaotic for $k > 1$. For even higher values of k, DERIVE struggles to produce a solution unless the value of n is reduced.

References

[1] Hollingdale S., *Makers of Mathematics*, Penguin (1989).

[2] Kaye G.W.C. and Laby T.H., *Tables of Physical and Chemical Constants*, 15th edition, Longmans (1986).

8

Dimensional Analysis

8.1 INTRODUCTION

As you work through this chapter you will find that the only mathematical skills required are the ability to handle indices and the ability to solve small systems of simultaneous linear algebraic equations; hardly sufficient to warrant the support of DERIVE!

Nevertheless, we have included a chapter on dimensional analysis because of its immense power in enabling us to model complex systems relatively easily in an algebraic way. As you will see, the technique provides an alternative to often very expensive or dangerous experimentation and provides the link between the real world systems and the use of scale models to analyze their behaviour.

The validation stage of some of the examples describes experimental work which you could undertake in order to generate some data which could then be analyzed using DERIVE. Despite this rather tenuous connection with DERIVE we felt that this was a more realistic approach to the topic rather than to force the use of DERIVE artificially to tackle the mathematics described in the first paragraph above.

The use of scale models raises some interesting questions. For example, if the wind flow around a proposed new tall building is under examination and the engineer makes a $1{:}n$ scale model of the building to test in a wind tunnel, should the wind speed also be scaled by the factor n? Similar considerations about proportionality would arise in developing a model of the crater size due to an explosion of a known magnitude (a full-scale experiment might be too dangerous) or the spread of a fire through a new shopping precinct (too expensive and too dangerous to conduct full-scale experiments).

In this chapter we shall consider a technique known as dimensional analysis, which provides information about the form of the mathematical relationships between the selected variables, and suggests ways in which they might be grouped together. The groupings provide a guide for the modeller in making predictions about the behaviour of the full size system. Fundamental to the technique are the concepts of geometric similarity and the dimension of a physical quantity, and so we address these issues next.

8.2 GEOMETRIC SIMILARITY

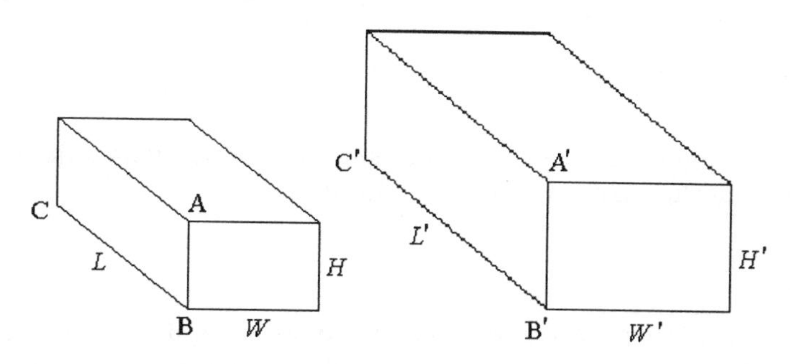

Figure 8.1 Geometrically similar boxes

Consider the two boxes shown in Fig. 8.1, one with sides of lengths L, W and H and the other of sides L', W' and H'. If one is simply a scale model of the other, then it is true that

$$\frac{L}{L'} = \frac{W}{W'} = \frac{H}{H'} = k$$

where k is a positive constant. A geometric consequence of this is that triangles such as ABC and A'B'C' are similar (the angles are in fact equal) – hence the name geometric similarity.

To consolidate these ideas you are advised to attempt the following exercises:

1. Show that the concept of geometric similarity (as expressed by the scale factor k above) leads to particularly simple expressions for the ratio of the surface areas (S and S') and volumes (V and V') of the two boxes.

   You should find that $S/S' = k^2$ while $V/V' = k^3$, the presence of the square and cube having an immediate intuitive appeal.

2. Deduce that $\dfrac{S}{L^2} = \dfrac{S'}{(L')^2} = k^2$ (constant) with a corresponding result in terms of volumes.

The side L of the rectangular box is referred to as a characteristic dimension (for a sphere one might choose the diameter, for a human being the height, for a bird the wingspan, and so on).

The value of the concept outlined above is evident in the following scenario. Suppose that when applied to a specific problem your modelling methodology has led to the mathematical model that your variable of interest (say y) is a function of length, surface area and volume, i.e.

$$y = f(L, S, V)$$

then the above arguments permit you to simplify your model from a function of several variables to a function of one variable (L), namely,

$$y = g(L)$$

The application of the ideas developed so far is illustrated in the following exercise about modelling the terminal velocity of a raindrop which is falling vertically:

1. Show that the downward motion of the raindrop is governed by the equation

$$ma = mg - R$$

where a is the downward acceleration of the raindrop, m is its mass, g is the acceleration due to gravity and R is the air resistance force.

2. Assume that R is proportional to the surface area (S) of the raindrop and the square of its speed (v) so that

$$R = c\,S\,v^2 \qquad (c \text{ a positive constant})$$

3. In terms of a characteristic dimension (L) of the raindrop deduce that

$$S \propto L^2 \text{ and } m \propto L^3$$

so that

$$S \propto m^{2/3}$$

4. Hence show that the terminal velocity (v_T) of the raindrop may be expressed in the form

$$v_T \propto m^{1/6}$$

5. Show that on impact with the ground the kinetic energy per unit area which must be converted into other energy is proportional to $m^{2/3}$ (assuming that the object has previously attained its terminal velocity).

8.3 DIMENSIONS

In the development of the various models in this text we have encountered a variety of different physical quantities such as time, displacement, velocity, acceleration, mass, force, energy, pressure, temperature, and so on. For example, *velocity = length/time*, *acceleration = velocity/time*, *force = mass × acceleration*, and so on. Regardless of the system of units adopted (conventionally the SI system), we repeatedly encounter the same basic 'building blocks' of

mass, length, time and temperature

These are often referred to as the base dimensions and are denoted respectively by

M, L, T and Θ

the latter symbol being an uppercase theta (to avoid confusion with T for time).

Hence we can write

$$\textit{The dimension of velocity} = L/T = LT^{-1}$$

which is abbreviated to

$$[\textit{velocity}] = LT^{-1}$$

the symbol [ ] being read as 'the dimension of' the quantity within the parentheses.

Similarly,

$$[\textit{force}] = MLT^{-2} \qquad [\textit{kinetic energy}] = ML^2T^{-2}$$

What are the dimensions of area, volume, density and pressure?

Some quantities are dimensionless such as angle in radian measure (as it is defined as the ratio of an arc length to the radius of the circle). Deduce the dimensions of angular velocity and angular acceleration. When a quantity is dimensionless its dimension is often expressed in the form $M^0L^0T^0$.

Constants which appear in equations can either have dimensions or be dimensionless. For example in the equation $A = \pi r^2$, relating the area to the radius r of a circle, π is dimensionless.

One of the gas equations of physics is $p = R\rho\Theta$ where p is the pressure, ρ is the gas density and Θ is the temperature. Show that the gas constant R has the dimensions $M^0L^2T^{-2}\Theta^{-1}$.

What are the dimensions of the universal gravitational constant, G?

8.4 DIMENSIONAL HOMOGENEITY

The only sensible equation is a consistent equation, i.e. an equation in which all terms have the same dimensions regardless of the system of units being employed.

The majority of equations encountered in practical problems contain sums of terms (for example $v^2 = u^2 + 2as$). In order to be able to add (or subtract) physical quantities they must have the same dimensions and then their sum (or difference) has these dimensions, too.

Confirm the dimensional homogeneity of the equation $s = ut + \frac{1}{2}at^2$, in which all terms have their usual meanings.

Dimensional homogeneity can be used to check the accuracy of an equation. For example the authors have seen the following equation in print:

$$p + \frac{1}{2}\rho u^2 = -mgz$$

in which all terms had their usual meanings.

A check for dimensional homogeneity gives

$$[p] = ML^{-1}T^{-2}, \quad \left[\frac{1}{2}\rho u^2\right] = ML^{-1}T^{-2}$$

the factor 1/2 being simply a dimensionless constant, but

$$[mgz] = ML^2T^{-2}$$

Something is obviously wrong somewhere, possibly with the *mgz* term. A check of other, related, technical literature revealed that the *mgz* term should in fact be ρgz, which was confirmed by dimensional considerations:

$$[\rho gz] = ML^{-1}T^{-2}$$

8.5 AND DERIVATIVES, TOO

Many of the models developed in this text have resulted in differential equations and this leads to the obvious question as to how the dimensions of derivative terms are handled. To answer this we simply appeal to the basic definition of a derivative as the limit of a ratio of changes in the two variables involved. For example, if p denotes pressure and z is a spatial displacement, then

$$\frac{dp}{dz} = \lim_{\Delta z \to 0} \frac{\Delta p}{\Delta z}$$

where Δz is a small change in z and Δp is the corresponding change in p. Now the changes Δp and Δz will have the same dimensions as p and z, respectively, and thus

$$\left[\frac{dp}{dz}\right] = [p]/[z]$$

$$= ML^{-1}T^{-2} \quad L^{-1}$$

$$= ML^{-2}T^{-2}$$

Note that partial derivatives can be treated in the same way.

Use the above approach to confirm that the dimensions of the derivative $\frac{dv}{dt}$ (where v is velocity and t is time) are indeed the dimensions of acceleration.

With the dimensional consideration of derivatives addressed, the concept of dimensional homogeneity can now be extended to include differential equations, as demonstrated by the following exercise.

Confirm the dimensional homogeneity of the wave equation

$$\frac{\partial^2 y}{\partial x^2} = \frac{1}{c^2} \frac{\partial^2 y}{\partial t^2}$$

where x and y are displacements, t is time and c is a constant having the dimensions of velocity.

The application of dimensional analysis to the modelling of real world problems is based on the assumption that the mathematical solution to the problem is given by a dimensionally homogeneous equation in terms of the selected variables.

The concept of dimensional homogeneity is formalized in Buckingham's theorem, a theorem central to the use of dimensional analysis in the solution of modelling problems. Buckingham's theorem states that an equation is dimensionally homogeneous if and only if it can be put into the form

$$f(\Pi_1, \Pi_2, \ldots, \Pi_n) = 0$$

where f is some function of n arguments and $\{\Pi_1, \Pi_2, \ldots, \Pi_n\}$ is a complete set of dimensionless products.

8.6 THE METHOD OF DIMENSIONAL ANALYSIS

The Problem

We shall illustrate the application of dimensional analysis to the solution of modelling problems by developing a model of the period (t seconds) of a simple pendulum.

Setting up a Model

Initial thoughts about relevant features produced a list as follows:

- the length of the pendulum (ℓ metres (m)) ,
- the mass of the pendulum bob (m kg) ,
- the acceleration due to gravity (g ms^{-2}) ,
- the initial angular displacement of the pendulum (θ rad) ,
- frictional forces at the point of suspension (F Newtons (N)) ,
- air resistance forces experienced by the bob (R Newtons (N)) .

Our initial model will be developed on the assumptions that

- the string is weightless – reasonable if the bob is significantly heavier than the string,
- the mass of the bob is concentrated at a point at the end of the string – enables a particle model to be developed,
- the resistive forces F and R will be ignored – enables a simple model to be developed initially.

Formulate a Mathematical Model

According to Buckingham's theorem we then find all dimensionless products amongst the variables t, ℓ, m, g, θ. Any such product must be of the form

$$t^a \; \ell^b \; m^c \; g^d \; \theta^e \qquad (8.1)$$

and thus has dimension

$$T^a \; L^b \; M^c \; (LT^{-2})^d \; (M^0 \; L^0 \; T^0)^e \; = M^c \; L^{b+d} \; T^{a-2d}$$

A product of the form (8.1) is therefore dimensionless if and only if

$$M^{c}\, L^{b+d}\, T^{a-2d} = M^{0}\, L^{0}\, T^{0}$$

whence

$$\left.\begin{aligned} c &= 0 \\ b+d &= 0 \\ a-2d &= 0 \end{aligned}\right\} \quad (8.2)$$

Solution of the Mathematical Model

We are now in a somewhat tricky position mathematically speaking as we have five unknowns (a, b, c, d and e) but only three equations in the system of (8.2). Solution gives $c = 0$, $a = 2d$, $b = -d$ where d is arbitrary as also is e (since it did not even appear in system (8.2)). Thus there is an infinity of solutions corresponding to the complete set of possible pairs of values of (d, e).

Since

$$(d, e) = d\,(1, 0) + e\,(0, 1)$$

then all dimensionless products can be obtained for the two cases

$$(d, e) = (1, 0) \quad \text{and} \quad (d, e) = (0, 1)$$

$(d, e) = (1, 0)$ gives the solution $a = 2$, $b = -1$, $c = 0$ while $(d, e) = (0, 1)$ gives the solution $a = b = c = 0$.

Thus the complete set of dimensionless products consists of

$$\Pi_1 = t^2\, \ell^{-1}\, m^0\, g^1\, \theta^0 = t^2 g/\ell$$
$$\Pi_2 = t^0\, \ell^0\, m^0\, g^0\, \theta^1 = \theta$$

Interpretation and Validation

Buckingham's theorem thus implies that there is a function f such that

$$f\left(\frac{t^2 g}{\ell}, \theta\right) = 0$$

Assuming that this equation can be solved for $t^2 g/\ell$ in terms of θ, then it follows that

$$t = k(\theta)\sqrt{\frac{\ell}{g}} \tag{8.3}$$

where k is some function of θ.

Dimensional analysis cannot give us any information about the function $k(\theta)$ and at this stage we would have to conduct some experimental work to establish its nature. (As you are probably aware, if the oscillations are restricted to being of small amplitude then $k(\theta) = 2\pi$).

Even this simple example shows the power of dimensional analysis: without performing any experiments we have shown that the period of the simple pendulum is independent of m and that its square is proportional to ℓ. For more complicated problems the potential savings in expensive experimental work are obvious.

Further Work

Next we investigate a revision of the above model by including the air resistance force experienced by the pendulum bob. The dimensionless products we now seek are therefore of the form

$$t^a \ell^b m^c g^d \theta^e R^f \tag{8.4}$$

Repeating the procedure detailed in the initial model, such products are dimensionless if and only if

$$\left.\begin{aligned} c + f &= 0 \\ b + d + f &= 0 \\ a - 2d - 2f &= 0 \end{aligned}\right\} \tag{8.5}$$

Once again we have an infinity of solutions as there are six unknowns ($a, b, \ldots, f$) but only three equations. We could solve the equations (8.5) for a, b, c to give

$$a = 2d + 2f, \; b = -d - f, \; c = -f$$

in terms of the arbitrary variables d, e and f. Since

$$(d, e, f) = d(1, 0, 0) + e(0, 1, 0) + f(0, 0, 1)$$

then all dimensionless products can be obtained for the three cases

$$(d, e, f) = (1, 0, 0) \ ; \ (0, 1, 0) \ ; \ (0, 0, 1)$$

These lead, respectively, to the dimensionless groupings

$$t^2 \ell^{-1} g \, , \, \theta \, , \, t^2 \ell^{-1} m^{-1} R$$

and Buckingham's theorem then implies the existence of a function g such that

$$g\left(\frac{t^2 g}{\ell}, \theta, \frac{t^2 R}{m\ell}\right) = 0 \tag{8.6}$$

Since t^2 appears in two of the arguments it is not necessarily a trivial matter to solve equation (8.6) and isolate t, the period. This difficulty could be overcome if the solutions of system (8.5) were chosen in such a way that t only appeared in one of the dimensionless groupings, and we investigate this in the first of the following exercises.

1. (a) Choose a, e and f as the arbitrary variables and show that the solution of equations (8.5) can then be expressed as

$$c = -f \, , \, b = -\frac{1}{2} a \, , \, d = \frac{1}{2} a - f$$

(b) Show that the dimensionless groupings corresponding to the three base (a, e, f) triplets are

$$\Pi_1 = t \sqrt{(g/\ell)} \, , \, \Pi_2 = \theta \, , \, \Pi_3 = R/m$$

(c) Use Buckingham's theorem to deduce that

$$t = \sqrt{\frac{\ell}{g}} \cdot K\left(\theta, \frac{R}{m}\right)$$

where K is some function of the arguments indicated.

2. A modelling investigation.

(a) Suppose that the air resistance force is given by the expression $R = cv^2$ where v is the velocity of the pendulum bob and c is a constant of proportionality, whose value depends on the shape of the pendulum bob. Use the method of dimensional analysis to show that the period of the pendulum may be expressed as

$$t = \sqrt{\frac{\ell}{g}} \cdot f\left(\theta, \frac{c\ell}{m}\right) \tag{8.7}$$

(b) Show that if the air resistance force is, instead, modelled as $R = \lambda v$, where λ is again a constant of proportionality whose value is dependent upon the shape of the bob, then the period of the pendulum may be expressed as

$$t = \sqrt{\frac{\ell}{g}} \cdot h\left(\theta, \frac{\lambda}{m}\sqrt{\frac{\ell}{g}}\right) \tag{8.8}$$

Hint: in both (a) and (b) above, work with the appropriate proportionality constant c ($= R/v^2$) or λ ($= R/v$) rather than the resistance force R itself.

Using only algebra and a variety of different modelling assumptions we have developed three different models for the period of a simple pendulum, represented by expressions (8.3), (8.7) and (8.8). The results help to focus our ideas about designing experiments which could be performed to validate the model(s). What do the three models predict?

Equation (8.3) predicts that for constant θ (the release position), then $t \propto \sqrt{\ell}$.

Equation (8.7) predicts that for θ and ℓ/m held constant while preserving the same drag characteristics (λ), then $t \propto \sqrt{\ell}$.

Equation (8.8) predicts that for θ and $\sqrt{\ell}/\mathrm{m}$ held constant while preserving the same drag characteristics (c), then $t \propto \sqrt{\ell}$.

In order to validate these models you could experiment with a pendulum consisting of a length of string and a hollow ball with a hole in it. The hole enables you to vary the mass of the ball without altering its drag characteristics. Don't forget to check out the special case mentioned earlier that for small angles (how small?) the period is independent of the release position.

8.7 RESISTANCE TO LOCOMOTION

Introduction

Many of the studies in this text have examined models for motion in resisting media. In this exercise we shall use dimensional analysis to develop a model of the resistance force experienced by cyclists, runners, motor vehicles, and so on.

The Problem

When racing, cyclists usually crouch low over their handlebars, presumably in an attempt to reduce the area they present to the wind. The wind has an effect on runners, too, and considerable savings of energy can be achieved by using another runner to shield oneself from a headwind. In the 1968 Olympic Games, held in Mexico City, the rarefied air helped the athletes to achieve greatly improved performances in the sprint events. Wherever they perform outdoors, the effects of air resistance are felt and it is therefore of interest to develop a model of this force.

Setting up a Model

Features on which the air resistance force R (Newtons) is likely to depend are

- the frontal area of the moving body (A m^2),
- the velocity of the body relative to the wind (V m s^{-1}),
- the air density (ρ kg m^{-3}).

The assumption on which we shall base our model is that the air resistance force can be represented in terms of powers of A, V and ρ as

$$R = k\,\rho^a A^b V^c \tag{8.9}$$

where k is a dimensionless constant.

Formulation and Solution of the Mathematical Model

In terms of the MLT system, consideration of the dimensions of equation (8.9) then gives

$$MLT^{-2} = (ML^{-3})^a\,(L^2)^b\,(LT^{-1})^c$$

Equating corresponding exponents on both sides of this equation gives a system of simultaneous equations whose solution is

$$a = 1\,,\quad b = 1\,,\quad c = 2$$

so that equation (8.9) becomes

$$R = k\,\rho A\,V^2 \tag{8.10}$$

Interpretation and Validation

Note that this is a resistance force model which we have used frequently throughout this text.

The result shows the benefit to a cyclist of crouching when cycling into a headwind, as A, and hence R, is reduced. A runner's 'shielding' strategy is similarly beneficial as it serves to reduce the wind velocity in the immediate vicinity of the runner. The mechanical benefit of running at altitude (where ρ is smaller than at sea level) is also apparent, although this is offset to some extent by the adverse physiological effects often associated with altitude.

Access to a wind tunnel is required to validate this model rigorously. If such access is available you could conduct an experiment to measure the resistance force R for different wind speeds V and objects of different cross-section A placed in the flow. A plot of R versus AV^2 will give a straight line if the model is obeyed.

8.8 A MUSICAL DIVERSION

Introduction

This exercise demonstrates how dimensional analysis can be used to develop a model for estimating the frequencies of the notes produced by different stringed musical instruments.

The Problem

Examination of different stringed instruments shows them to have strings of different lengths, different thicknesses and different tensions. Develop a model to predict how these factors influence the frequency of the notes produced by each instrument.

Setting up a Model

As suggested in the problem outline the features likely to affect the frequency w (Hz) of a note are

- the length of the string (ℓ metres (m)) ,
- the thickness of the string (radius r metres (m)) ,
- the material of the string (typified by its density) ,
- the tension in the string (F Newtons (N)) .

We shall assume the following:

- The vibration of the string is not subject to any damping force.
- Any elastic properties of the string are ignored.
- The thickness and material of the string are represented by a single parameter – the mass per unit length (ρ kg m^{-1}) .

Formulation and Solution of the Mathematical Model

Following Buckingham's method we then find all dimensionless products among the variables w, ℓ, ρ and F. Any such product must be of the form

$$w^a \ell^b F^c \rho^d$$

Show that the method of dimensional analysis leads to the equations

$$\begin{aligned} c + d &= 0 \\ b + c - d &= 0 \\ a + 2c &= 0 \end{aligned}$$

If d is chosen as the arbitrary variable deduce that there is one dimensionless product

$$\Pi_1 = \frac{w^2 \ell^2 \rho}{F}$$

According to Buckingham's theorem there exists a function f such that

$$f(\Pi_1) = 0$$

LIVERPOOL JOHN MOORES UNIVERSITY
LEARNING SERVICES

Hence $\Pi_1 = \text{constant}$, which implies

$$w \propto \frac{1}{\ell}\sqrt{\frac{F}{\rho}} \tag{8.11}$$

Interpretation

Equation (8.11) implies that for a given string with F held fixed, $w\ell$ is constant. Similarly, for a fixed length of a given string $w/\sqrt{F}$ is constant. Work by Oberndorf on the analysis of a similar model, reported in Bender [1], indicates that although equation (8.11) implies that $w\sqrt{\rho}$ is constant for given ℓ and F, this is in fact not the case, especially for very thin strings and very high tensions. Can you suggest why?

Validation

Design an experiment to investigate the above assertions by investigating a stringed instrument of your choice.

Further Work

If you have the necessary background knowledge of the properties of materials, investigate how your model would be modified by the further assumptions that the material of the string is elastic and isotropic.

8.9 FLOW THROUGH PIPES

Introduction

The flow of fluids through pipes is an important topic in disciplines ranging from engineering (e.g. the flow of hot water through heating/cooling systems) to haematology (e.g. the flow of blood through arteries).

The flow can be either turbulent or laminar; in the latter case the fluid moves steadily along the pipe in an 'orderly manner' and thus represents the desired situation rather than the irregular motion which is characteristic of turbulent flow.

In the case of laminar flow it would be useful to have an expression for the volume rate of flow through a pipe.

The Problem

Develop a model of laminar fluid flow through a pipe and use it to obtain an expression for the volume rate of flow in terms of the parameters of the pipe system.

Setting up a Model

Features thought likely to be of relevance to the problem are

- the flow rate through the pipe (q m^{3}s^{-1}) ,
- the length of the pipe (ℓ m) ,
- the radius of the pipe (r m) ,
- the viscosity of the fluid (μ Nm^{-2}s) ,
- the pressure gradient along the pipe (p/ℓ Nm^{-3}) .

You may wonder why we have not included fluid velocity in this features list. Initially, we did but then realized that the velocity was caused by the pressure gradient and hence velocity is implicitly included in the above list.

Formulation of a Mathematical Model

With the features for inclusion in our model identified, we turn to the modelling assumptions. We shall assume that the model can be expressed in terms of powers of the defined features in the form

$$q^{a}\, r^{b}\, \mu^{c} \left(\frac{p}{\ell}\right)^{d} \tag{8.12}$$

as required by the method of dimensional analysis.

Before investigating the solution of this model, a few words about viscosity may be appropriate for readers without an engineering background. Viscosity is simply a measure of the resistance to flow of a fluid. For example, water flows very easily but treacle does not; the viscosity rating of treacle is higher than that for water. Lubricants are classified according to their viscosity.

Solution of the Mathematical Model

The dimension of the product above is

$$(L^{3}T^{-1})^{a}(L)^{b}(ML^{-1}T^{-1})^{c}(ML^{-2}T^{-2})^{d} = M^{c+d}\,L^{3a+b-c-2d}\,T^{-a-c-2d}$$

The remainder of the solution is presented as a directed exercise for the reader.

Write down the resulting three equations in a, b, c and d in order that the product is dimensionless. By choosing d as the arbitrary variable show that their solution is

$$a = -1\,,\ b = 4,\ c = -1\ \text{and}\ d = 1$$

so that the appropriate dimensionless quantity is

$$\Pi = q^{-1}\, r^{4}\, \mu^{-1} \left(\frac{p}{\ell}\right)^{1}$$

Use Buckingham's theorem to deduce that

$$q^{-1}\, r^{4}\, \mu^{-1} \left(\frac{p}{\ell}\right)^{1} = \text{constant}$$

and hence that

$$q \propto \frac{pr^{4}}{\mu \ell} \tag{8.13}$$

Interpretation

Expression (8.13) shows that for a given size of pipe and a specific fluid (i.e. r and μ are fixed), the volume flow rate is proportional to the pressure. Again, for fixed values of the pressure, radius and viscosity the volume flow rate is inversely proportional to the length of the pipe. What further interpretations can you make?

Validation

Fig. 8.2 shows an apparatus which could be used to investigate the validity of the above interpretations. The overflow pipe CD ensures that the end A of the horizontal pipe is at a constant depth below the free fluid surface, giving rise to the name 'a constant head apparatus'.

The pressure at A = *atmospheric pressure* + $\rho g z$ where ρ is the fluid density and pressure at B = *atmospheric pressure*. Hence the pressure gradient along the pipe AB is $\rho g z/\ell$.

For fixed values of p, μ and ℓ measure the volume flow rate for pipes of different radius and then use DERIVE to investigate the q versus r^4 fit and hence establish the constant of proportionality for expression (8.13). Confirm your result by investigating the q versus $1/\ell$ fit for fixed values of p, μ and r using a range of pipes of different lengths and DERIVE's FIT function to analyze the data.

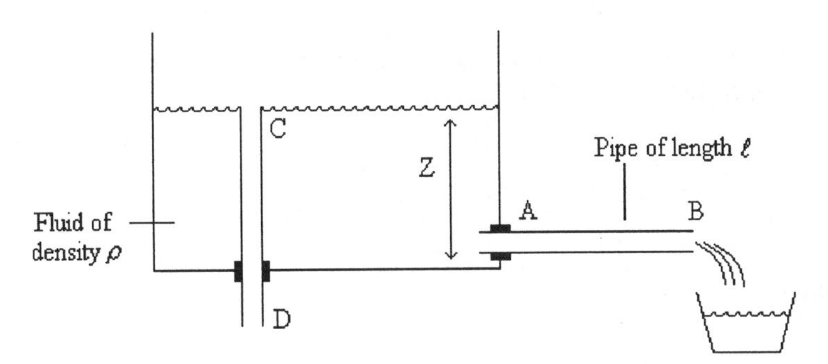

Figure 8.2 Apparatus for validation of model of flow through pipes

Finally, you may care to investigate Poiseuille's formula in undergraduate engineering fluids textbooks.

8.10 SIMILITUDE

Finally, we return to the question relating to scale models which was posed in the introduction to this chapter. Dimensional analysis helps to solve the problem of how to scale the parameters of a system so that accurate predictions can be made about the real problem by analyzing a scale model (or prototype).

Before looking in detail at this problem of scaling we consider one final application on dimensional analysis, as it serves to introduce a dimensionless product which is fundamental to the application of the principles of similitude in fluid mechanics.

Assume that the resistance force R experienced by a smooth sphere as it moves through a fluid depends on the velocity v of the sphere, its diameter d, the density and the viscosity of the fluid ρ and μ respectively. Determine the simplest form of the dependence by carrying out the following.

Show that selecting dimensionless products of the form

$$R^a\ v^b\ d^c\ r^e\ \mu^f$$

leads to the requirement that

$$a + e + f = 0$$
$$a + b + c - 3e - f = 0$$
$$2a + b + f = 0$$

Select a and e as the two arbitrary variables, solve the above equations and apply Buckingham's theorem to show that

$$f\left(\frac{R}{vd\mu}, \frac{vd\rho}{\mu}\right) = 0$$

and hence that

$$R = \mu vd\ F\left(\frac{\rho vd}{\mu}\right)$$

where $\rho vd/\mu$ is a dimensionless product and F is a function of this argument.

The quantity $\rho vd/\mu$ is known as Reynolds number (after Osborne Reynolds 1842 –1912) and is denoted by Re. It is usually defined in terms of a characteristic length (ℓ) involved in the problem (in our example it was the diameter d) so that

$$Re = \frac{\rho v \ell}{\mu}$$

It is of fundamental importance in aerodynamics and fluid mechanics.

The basic idea for determining how the scaling for a model should be performed comes from Buckingham's theorem. Suppose the full size problem can be described in terms of the dimensionless variables $\Pi_1, \Pi_2, \ldots, \Pi_n$ so that

$$f(\Pi_1, \Pi_2, \ldots, \Pi_n) = 0$$

and that the desired dependent variable only occurs in one of them, say Π_n, so that

$$\Pi_n = F(\Pi_1, \Pi_2, \ldots, \Pi_{n-1})$$

In order that the solution for the model and the prototype are the same, then it is sufficient that the value of each of the dimensionless products $\Pi_1, \Pi_2, \ldots, \Pi_{n-1}$ is the same for the model and the prototype.

For example, suppose we are developing a hull design for a ship. Then if v_p is the design speed of the ship through the water, ρ_p and μ_p are the density and viscosity of sea water and ℓ is a characteristic dimension of the vessel (perhaps the waterline length), the Reynolds number associated with the prototype is

$$Re = \frac{\rho_p \, v_p \, \ell}{\mu_p}$$

Obviously, it is undesirable to build a full size ship and then find that it does not perform as expected so, instead, a scale model is produced. Let v_m, ℓ_m, ρ_m and μ_m be the corresponding parameter values for the scale model experiments. Then we require

$$\frac{\rho_p \, v_p \, \ell}{\mu_p} = \frac{\rho_m \, v_m \, \ell_m}{\mu_m} \tag{8.14}$$

Thus if we make a 1:20 scale model of the ship we could use the same fluid in the model tank if we increased the flow velocity by a factor of 20. If this were considered too great a velocity we could scale by a lesser amount, say λ, and use a different fluid in the model tank such that

$$\frac{\lambda \rho_m}{20 \, \mu_m} = \frac{\rho_p}{\mu_p}$$

By judicious choice of the model parameters we can therefore ensure that condition (8.14) is satisfied so that the performance of the model can be considered to reflect the performance of the prototype. We included the word 'judicious' because it is necessary to be careful when scaling so that effects which are negligible for the prototype do not become significant for the model. Can you think of such an effect in this example?

Once again, this highlights the need for the modeller to seek expert advice sometimes and/or listen to those with experience of the problem being modelled. However sophisticated you may believe your model to be, it is worthless if it does not address the client's problem!

Reference

[1] Bender E.A., *An Introduction to Mathematical Modelling*, Wiley, New York (1978).

9

A Modelling Miscellany

9.1 INTRODUCTION

This chapter contains a selection of modelling problem statements, some of which have starter hints provided. They are grouped together in this chapter rather than distributed through the earlier chapters so that you can approach each with an open mathematical mind rather than be influenced by the mathematical content of the chapter in which it is presented.

The problem statements presented offer you the opportunity to develop your own mathematical models from scratch, perhaps collect your own data and use DERIVE as your mathematical assistant in obtaining the solution of the models you develop.

9.2 WAITING FOR AN ELEVATOR

A common excuse made by students (and lecturers!) for lateness at lectures is the delay caused by waiting for elevators. At the authors' institution, one particular building has eight floors (i.e. a ground floor numbered 0 and then floors numbered 1–7 inclusive) and is served by two independently operating elevators situated side-by-side. Each elevator holds a maximum of ten average-sized persons. A nearby stairwell offers an alternative means of reaching a given floor.

Every hour, just prior to the start of lectures, students leave the coffee room on the ground floor and have to decide whether to catch an elevator or climb the stairs. After numerous complaints about waiting time, the elevators are being modernized to try to improve the efficiency of their operation and a model is needed in order to attempt to establish the optimum strategy for their operation. For example, should elevators only stop at alternate floors or should one elevator only be allowed to stop at floors 5,6 and 7 after leaving the ground, etc.?

For the above case, students leaving the coffee room are equally likely to be heading for any one of the seven floors above. Timings taken show, that when in motion, an elevator takes an average of three seconds to rise one floor and, when stationary, the elevator doors take four seconds to move from closed to fully open. Determine a strategy which will ensure that as few people as possible are late for lectures.

Comment

Readers can attempt to model the above situation or adapt the problem to reflect the operation of elevators in their own locality. All data considered necessary to formulate a model (or models) and to evaluate the usefulness of such models can then be gathered.

9.3 GROWING TOMATOES IN GROW-BAGS

Many keen amateur gardeners grow their own tomatoes by planting the small tomato plants in grow-bags which can be laid down in greenhouses or on patios and then removed once the tomatoes have been harvested.

Grow-bags come in various sizes, but the most common are 1 m × 40 cms in cross-section and the user is normally advised to cut out three rectangular holes of size 30 cm × 25 cm. and to put one tomato plant in each hole for best results. However, it is very tempting to try to 'squeeze' some extra tomato plants into a grow-bag either by putting two plants into each hole (i.e. six plants) or to cut out four smaller rectangles instead of three and put one plant in each (i.e. four plants). Since tomato plants are relatively inexpensive, a few 'failures' (due to overcrowding of the plants) do not represent a major financial loss.

Clearly, the more plants there are in a confined space, the greater the competition for the light, water, etc., needed for healthy growth. A simple model would be helpful for gardeners to try and convince them that an average yield from (say) six plants is not as good as an optimum yield from three plants and that perhaps the best course of action is to follow the advice given by the grow-bag suppliers.

Comment

Readers may decide that they need expert advice on growth-rates, etc., from an expert gardener or biologist. Perhaps a more appealing move is to try to formulate a model based on your own experiments with growing tomatoes in this way.

9.4 REPAYMENT VERSUS ENDOWMENT MORTGAGES

Financial advisors, banks, building societies, etc., are forever formulating new schemes for repaying a mortgage on a property. Their schemes are often targeted at particular groups of borrowers such as first-time buyers, people paying higher rates of tax, etc.

Two of the most common methods of settling a mortgage are:

(i) a repayment mortgage, where the total debt over the loan period, including interest, is calculated and the amount repayable per month determined to discharge the debt by the end of the loan period,

(ii) an endowment mortgage, in which the amount owed is met by payment into an endowment policy which matures at the end of the loan period into a sum that hopefully covers the mortgage debt and an additional lump-sum benefit to the mortgagee.

Both systems are affected by variations in the interest rate. The general advice often given is that when interest rates are high it is beneficial to take out an endowment mortgage for as long as possible, but the situation for low interest rates or the influence of the length of the loan is not always clear. In recent times, for example, with relatively low interest rates, some financial advisors have been accused of unfairly favouring endowment mortgages and some building societies have adopted a practice of only offering repayment mortgages to certain groups of borrowers.

A model is needed to enable different groups of borrowers to judge for themselves which method is likely to be more beneficial.

Comment

Relevant case-study data are often available from people with fully paid-up mortgages and your model should show how well they were advised when they took their mortgage out – they will probably not wish to know if your model shows they were ill advised!

9.5 COLLECTING FOOTBALL STICKERS

Most young boys like to collect such items as football cards or stickers. The current trend is for the sticker book to be offered free of charge with a national daily newspaper and for the stickers to be on sale. One such example meant that 500 individual stickers were offered for collection in a series where the stickers could be bought in packets of six for 20p. The company distributing the stickers guaranteed that all the stickers were produced in equal numbers (but does this mean that they were equally available in all regions of the country?) and that in a packet of six stickers, no two stickers were the same. The distributing company also gave the collector the opportunity to send off for thirty stickers of the collector's choice at 6p per sticker. They also encouraged the swapping of stickers among collectors.

One of the authors, with a son who collects such stickers, would be as interested as most parents in the development of a model to estimate the likely cost of collecting the above series of stickers. More generally, it would be interesting to know whether the distributing company was optimizing its profit by choosing 500 items and six in a packet.

Comment

This problem is similar to that of the 'cigarette card collector' problem which is discussed as an example of a negative binomial distribution in [1]. However, the detailed probability theory underlying this problem is somewhat complicated and the reader is encouraged to start by developing simpler models. Attempting to model the swapping arrangements can also be difficult, but it may be possible to develop best and worst case estimates.

9.6 TRIATHLONS

The most common distances used in triathlon competitions are 1.5 km swimming, 10 km running and 40 km cycling. The question arises as to whether these values are used for the convenience of the organizers of triathlons or do they correspond to proportions of the three activities which imply that the winner is the best triathlete?

Develop a model which could be used to determine whether the proportions corresponding to the above distances represent the optimum values for determining the best triathlete.

Points for you to ponder are as follows:

- What criteria should be used for comparing the different components?
- Do current triathlon results indicate that any component(s) is/are more important than any other?

Apart from the specialist triathlon magazines, sports physiology text books are likely to provide relevant data, see, for example, [2].

9.7 TRAFFIC FLOW

Two of the major road routes into the City of Liverpool involve travel through a tunnel.

In the morning and evening rush hours the traffic flow can be very congested and delays sometimes occur resulting in frustation (leading to bad driving) and economic loss to businesses.

The traffic manager of the Tunnel Authority realizes that a policy is needed which will maximize the vehicle flow rate at these busy times while having due regard to the safety of the drivers involved.

You are consulted by the Authority and asked to recommend a vehicle speed and separation which will achieve this. Develop a mathematical model which could be used to investigate this problem. What advice would you give to the Tunnel Authority?

Comment

The model could be developed to consider

- the effect of having toll collecting booths (manned and automatic) at the entrances to the tunnel,

- the effect of any gradients within the tunnel (e.g. slow lorries going uphill),
- the effect of having city streets at one end of the tunnel and a motorway at the other, ...

9.8 POPULATION DRIFT

We read a lot these days about the decline in the population of major cities. There is evidence from population censuses and surveys to indicate that the probability that a person who currently lives in the city will still live in the city in ten years is 0.8 and that the probabitity that someone who currently lives outside the city will move into the city in ten years is 0.1 .

If the current city and suburban populations are 65 000 and 75 000, respectively, investigate whether a stable population distribution is possible.

How sensitive is your conclusion to changes in the relocation probability estimates and to the current populations?

9.9 QUEUING

Everyone has to queue at some time. It is rather frustrating to be stuck in a queue and one often hears such comments as 'You would have thought they would be able to cope by now, it's always like this when I come in.'

Why do queues develop?

Develop a model of queuing and use it to explain some of the experiences we share in queues. In order to develop a simple model initially, consider the case of one server, regular arrivals (say one per minute) and random service times.

The ideas can be extended further to consider

- a model for the patients in a doctor's waiting room with or without an appointments system,
- a comparison of a single queue with many servers (such as a bank or post office) with one queue per server.

9.10 SKI JUMPING

Nordic ski jumping involves travelling down an inclined slope, taking off from its horizontal end and subsequently travelling through the air to land on a lower, inclined slope, as shown in Fig. 9.1.

Develop a mathematical model which could be used to determine the skier's trajectory and, in particular, the distance of the landing point from the foot of the take-off tower, assumed to be 10 m vertically below the take-off point. *

Typical data values for this activity may be assumed to be as follows:

- Aerodynamic drag coefficient, C_D , is 0.8.
- Aerodynamic lift coefficient, C_L , is 0.4.
- Cross-sectional area, A m^2, presented to the air by the skier while in flight is 0.7 m^2 .
- Typical take off velocity is 15 m s^{-1}.
- Inclination of landing slope to the horizontal is 45°.

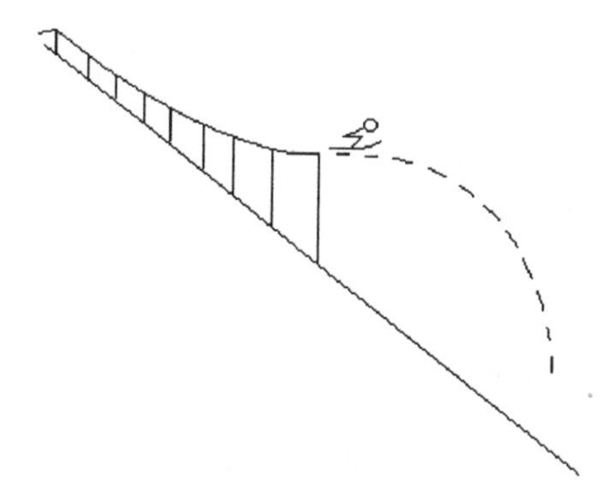

Figure 9.1 Ski jumping trajectory

9.11 ROASTING A TURKEY

A turkey (like any other piece of meat) is considered to be cooked when its internal temperature reaches a certain value. The internal temperature required depends on the type of meat being cooked.

Explain why the cooking time (t) is likely to depend on the difference in temperature ($\delta\theta_m$) between the raw meat and the oven temperature, the difference in temperature ($\delta\theta_c$) between the cooked meat and the oven, some measure of the size (ℓ) of the turkey and some measure (k) of the ability of the meat to conduct heat. (This last measure is generally taken to be the thermal conductivity of the meat.)

Develop a mathematical model to determine the dependence of the cooking time on the weight for similar pieces of meat. Use the model to compare the predicted results with the cooking instructions presented below

* This model could be extended to include the influence of the length and inclination of the take-off slope, waxing of the skis, etc., to convince yourself that this is indeed a typical take-off velocity.

> For turkeys up to twelve pounds in weight, the cooking time is fifteen minutes per pound plus an extra fifteen minutes, then add twelve minutes per pound up to twenty one pounds.

A rule of thumb quoted to the authors for the roasting of turkeys was simply to allow twenty minutes per pound. How does this rule compare? Can you suggest a reason why so many cooks overcook a turkey?

9.12 FIRE ALARMS

We have probably all experienced practice fire alarms and the subsequent evacuation of our school, university, factory or office.

The evacuation of the premises must be achieved in as short a time as possible consistent with there being no panic.

Choose a building that is familiar to you and develop a model which could be used to predict the time required to evacuate it completely. How do your results compare with the actual evacuation time?

9.13 HOSPITAL CORRIDORS

Have you ever wondered why hospital corridors are so wide? In answer to our question, our local hospital explained that it was mainly so that trolley beds could be easily moved around the hospital without the need for any complicated manoeuvres at right angle corners.

Investigate how narrow a corridor can be and yet permit a typical bed to negotiate a right angle corner.

9.14 RANKING SPORTS PERFORMANCES

In sports such as ski jumping, ice skating, gymnastics, etc., the performances of the competitors are in effect ranked by the judges. In many cases, the component marks awarded by the judges are combined to produce a final mark. In ice skating, for example, the skaters are awarded marks for technical merit and artistic impression for each of the compulsory figures and free skating programmes. In ski jumping, the skiers are given a combined mark based on the distance jumped and the style of the jump (the latter awarded by judges).

For an appropriate sport of your choice, investigate the way the marks are used to rank performance. Does the ranking scheme take into account any variability in judging (e.g. by discarding the highest and lowest marks awarded)? Are the final placings based on a collection of component marks or on the placings in the component parts of the performance? Investigate various models for ranking strategies. Can you suggest any scheme for your own chosen sport which seems fairer than the one currently used?

9.15 THE COST OF LEISURE

One of the easiest ways of spending money quickly is to take children to a theme park that offers 'white knuckle' rides. Consider two examples located in England.

The first is Blackpool Pleasure Beach, where the customer buys as many tickets as he/she wishes from a central ticket booth and exchanges the appropriate ticket for each ride taken. Typically, rides are £2 each and tickets can be bought separately or in batches with a discount, e.g. twelve tickets for £20. Admission to the Pleasure Beach complex is free.

The second is Alton Towers, where visitors pay to enter the theme park (typically £15 per adult, £11 per child with various discounts possible) and thereafter all rides are free.

Which offers the better value for money for the customer (who could be a single child, single adult or a family of adults and children)? Which optimizes the income for the owners of the theme park? Which is likely to lead to more queuing for rides? Is the length of a ride likely to be a factor when deciding pricing policy? Which park is likely to be affected more by adverse weather conditions (if at all)?

Develop a model which will answer these and any other questions you feel are relevant to a parent having to pay for the family's entertainment.

9.16 THE SPREAD OF DISEASE

The spread of disease among the human and animal populations is an important issue for us all. Models are needed to predict the spread of a disease so that effective patterns of treatment can be attempted or so that social practices can be changed to minimize the spread of disease. At this time, one could consider such cases as the spread of AIDS, the outbreak and spread of measles, 'mad cow' disease in cattle, and many more.

There are numerous examples in the literature of the development of models of the spread of disease, ranging from those using fairly simple mathematics to those involving advanced mathematical concepts. A readable example of modelling the spread of sexually transmitted diseases can be found in [3].

In developing your model, you may need to consider the following points, depending upon the disease being studied:

- Can the disease be treated once a person is infected and, if so, how successful is the treatment?
- If infected and then cured, does having had the disease ensure immunity from that disease again?

- What are the time scales involved? Full AIDS might take years to develop whereas the incubation period for measles is only a few days.
- Does the disease affect all the individuals equally or are some groups in the population more at risk, e.g. men versus women, young children and pensioners, poorer sections of the community, etc.
- and many more.

Comment

You are advised to start with a simple model that makes many assumptions and includes only those features considered most significant. The model can then be increased in complexity. Parameter values needed to specify the model, e.g. percentage of successful treatments of a disease, known cases of a disease in a given year, etc., can usually be found in official government statistical reports. Such data can also be used to confirm or deny the value of your model's predictions for known cases. You may need to seek expert advice about the nature of the disease you are modelling.

References

[1] David F.N. and Barton D.E., *Combinatorial Chance*, Griffin, London (1962).

[2] Reilly T., Secher N., Snell P. and Williams C., *Physiology of Sports*, E. and F.N. Spon, London (1990).

[3] Smith R., 'Sexually transmitted diseases among male homosexuals.', *Teaching of Mathematics and its Applications*, **8** (3), 109–114 (1989).

10

Learning Modelling – things that get in the way

10.1 INTRODUCTION

The past few years have seen a dramatic change in the mathematical skills base of many entrants to university undergraduate mathematics courses. In many cases, the enhanced problem-solving skills have been acquired at the expense of familiarity and facility with some of the basic mathematical techniques on which modelling tutors have been able to rely on in the past.

For example, while most students can quote the equation of a straight line and may thus be able to develop linear models, a majority are unaware of the related linearizing effect which logarithms may have on a set of data. Similarly, students will almost certainly have encountered the exponential function, but faced with a set of graphs of different exponential functions, they may well be unable to identify which decays/grows at the greatest/smallest rate. Tutors can doubtless think of examples of their own. A modelling analogy could be a brick wall for which the foundations are in place but for which much of the detailed brickwork is missing.

Our experience is that prior to embarking on a modelling course many students would benefit from some introductory lectures to re-enforce the behaviour of, for example,

exponential functions – growth and decay, with and without limit,
periodic functions – oscillations, vibrations, environmental and physical occurrences. Damping effects,
linear equations – and quasi linear forms, e.g. the equation $y = ax + bx^2$ expressed in linear form as $y/x = a + bx$,
log-log and semi-log plots – their effect on selected data sets,
input-output models and approximations – calculus and small increments approach,
translation into mathematics – extracting the mathematics from the text of a problem.

The time involved need not be excessive as much progress can be made using an investigative approach supported with access to a computer-algebra package. Any time penalties incurred are more than compensated for by the enhanced mathematical confidence and appreciation of the utility of the above functions and techniques brought to their modelling.

10.2 DERIVE ACTIVITIES

The following list of investigations and activities is indicative of how DERIVE can be used to improve rapidly the students' understanding of many of the functions previously mentioned.

1. By plotting graphs of e^{-t} and e^{-2t}, decide which decays faster. (Alternatively, present both graphs and ask which is which.). Conjecture the link between the exponent and the rate of decay and confirm/refute your conjecture by examining e^{-3t}, e^{-4t}, etc.
2. Plot graphs of e^{2t} , $e^{0.5t}$ and comment upon the growth rates. Conjecture and investigate as above.
3. Plot graphs of the functions $1-e^{-t}$, $1-e^{-2t}$ and $\frac{1}{1+e^{-t}}$. Comment on the behaviour and growth rates. Conjecture a 'rule' relating to their behaviour. (These expressions are typical of growth rate curves.)
4. Conjecture what the shape of the graph $y = te^{-t}$ might be. Confirm or refute your conjecture by plotting graphs of the functions $y = t$, $y= e^{-t}$ and $y = te^{-t}$. Notice that after an initial period of growth (for small values of t) the function decreases as the power of the exponential function prevails. (This phenomenon is sometimes observed in studies of resource exploitation.)
5. Consider the quadratic function at^2+bt+c. By selecting different triplets of values (a,b,c) deduce that (i) the curve is concave upwards if $a > 0$ and concave downwards if $a < 0$, (ii) the maximum or minimum value occurs at $t = -b/2a$ and (iii) that the expression may have one, two or no roots as evidenced by the number of intersections with the t axis.
6. Extract the vehicle stopping distance data from the Highway Code and observe that they fit the relationship

   $$d = v + 0.05\, v^2$$

   where d (feet) denotes the stopping distance and v (mph) denotes the vehicle speed. By plotting a graph of d/v against v confirm the above relationship by measuring the slope and intercept of the resulting straight line.

   Similar linearizations can often be effected for other equations. For example, an equation of the form

   $$y = ax\, /\, (b + x)$$

   can be linearized by plotting $1/y$ against $1/x$.

7. Plot graphs of 3 sin(t), sin($3t$) and sin($3+t$) and notice the effect of the 3 as it 'moves around' the expressions. Investigate the function a sin($bt+c$) for different values of a, b and c. In particular, note the effect of the parameter b on the period of the oscillations and the effect of c on the zeros of the function.
8. Plot graphs of the functions $e^{-0.1t}$ and sin($3t$) on the same axes, together with their product $e^{-0.1t}\sin(3t)$. Describe the behaviour of the product function and suggest an area of application in which such an expression would be appropriate.
9. Make a conjecture as to the behaviour of the function t sin($2t$) and confirm/refute it by plotting the expression. Suggest an area of application in which such an expression might be appropriate.

The utility of log–log or semi–log graph paper can be vividly demonstrated by presenting students with sets of data which follow either a power law ($y = ax^b$) or an exponential law ($y = ae^{bx}$). A graph of the raw data produces a curve in each case. Students can then be invited to explore the effects of plotting other graphs involving the logarithm of one or both of the original variables to ultimately produce a straight line. Once the slope and intercept of the line has been measured, the values of the parameters a and b can be deduced. The DERIVE function FIT is useful here.

Many of the models which involve rates of change can be developed by applying the so-called input-output principle to the problem. This principle states that consequent upon a small change in the independent variable then

$$\textit{net rate of change} = \textit{rate of input} - \textit{rate of output}$$

For example, consider a rectangular tank of cross-section A in which the depth of water at time t is h. Then if the tank is being supplied at a volume flow rate q_i and drained at a volume flow rate q_o then in a time δt, application of the above principle gives

$$A\delta h = q_i\,\delta t - q_o\,\delta t$$

from which a differential equation model can be developed. Our experience is that while students are quite competent at calculus technique-bashing, the above first-principles approach needs considerable re-enforcement.

10.3 TRANSLATION INTO MATHEMATICS

The final topic which we re-enforce is 'translation into mathematics'. By this we mean the extraction of relevant information from a problem statement expressed in words and its subsequent formulation into mathematical notation. Examples such as the following, based on population modelling, have been found to be useful.

Read the following passage, sketch a graph which represents the behaviour described and express the passage in mathematical terms.

> A rare species of animal is to be introduced into a new environment with initially only a few animals (*A*). Their numbers grow slowly to begin with (*B*) but once established they enjoy a period of faster, steady growth (*C*) before the growth rate declines as the species uses up the space and food resources and reaches an equilibrium level (*D*). The effects of immigration and emigration may be ignored in developing the model. Can you justify that this is reasonable?

This may be summarised graphically as shown in Figure 10.1.

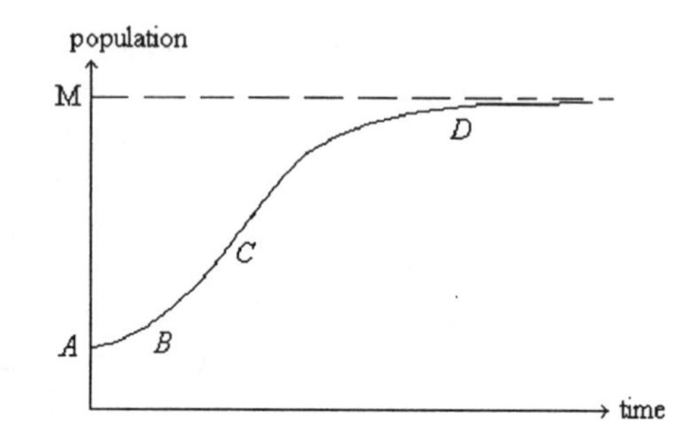

Figure 10.1 Population growth to a limiting value

Mathematically the situation may be modelled as

rate of growth of population	$\propto$	population size	and	proximity of population size to equilibrium level

$$\frac{dP}{dt} = kP\cdot(M-P) ,$$

where k is a positive constant.

A second example used in this 'translation section' is to consider the statement 'y decreases as x increases' and to invite students to suggest plausible mathematical expressions which represent this. They can then be asked to indicate which expression they would adopt as an initial mathematical model and why. Data sets can easily be developed to support the linear model or an inverse proportional model; the possibilities are endless, but with investigations and class discussions as outlined in this chapter, mathematical modelling students can be guided through the myriad of mathematical functions and given some pointers as to which functions may be appropriate for a particular problem.

DERIVE Functions and Utilities

Index